# ACTION RECOGNITION IN THE VISUAL PERIPHERY

Dissertation

zur Erlangung des Grades eines

Doktors der Naturwissenschaften

der Mathematisch-Naturwissenschaftlichen Fakultät

und der Medizinischen Fakultät

der Eberhard-Karls-Universität Tübingen

vorgelegt von:

Laura Fademrecht

aus Reutlingen, Deutschland

February 2017

Bibliografische Information der Deutschen Nationalbibliothek

Die Deutsche Nationalbibliothek verzeichnet diese Publikation in der
Deutschen Nationalbibliografie; detaillierte bibliografische Daten sind
im Internet über http://dnb.d-nb.de abrufbar.

ISBN 978-3-8325-4448-5

Logos Verlag Berlin GmbH
Comeniushof, Gubener Str. 47,
10243 Berlin
Tel.: +49 (0)30 42 85 10 90
Fax: +49 (0)30 42 85 10 92
INTERNET: http://www.logos-verlag.de

Tag der mündlichen Prüfung: 31. Januar 2017


Dekan der Math.-Nat. Fakultät: Prof. Dr. Wolfgang Rosenstiel

Dekan der Medizinischen Fakultät: Prof. Dr. Ingo B. Autenrieth


1. Berichterstatter: Prof. Dr. Heinrich H. Bülthoff

2. Berichterstatter: Prof. Dr. Martin A. Giese


Prüfungskommission: Prof. Dr. Heinrich H. Bülthoff

Prof. Dr. Martin A. Giese

Dr. Andreas Bartels

Dr. Tobias Meilinger

# Declaration

I hereby declare that I have produced the work entitled: "*Action Recognition in the Visual Periphery*", submitted for the award of a doctorate, on my own (without external help), have used only the sources and aids indicated and have marked passages included from other works, whether verbatim or in content, as such. I swear upon oath that these statements are true and that I have not concealed anything. I am aware that making a false declaration under oath is punishable by a term of imprisonment of up to three years or by a fine.


Laura Fademrecht


Laura Fademrecht                    February 17

# SUMMARY

Humans are social beings that interact with others in their surroundings. In a public space, for example on a train platform, one can observe the wide array of social actions humans express in their daily lives. There are for instance people hugging each other, waving to one another or shaking hands. A large part of our social behavior consists of carrying out such social actions and the recognition of those actions facilitates our interactions with other people. Therefore, action recognition has become more and more popular as a research topic over the years. Actions do not only appear at our point of fixation but also in the peripheral visual field. The current Ph.D. thesis aims at understanding action recognition in the human central and peripheral vision. To this end, action recognition processes have been investigated under more naturalistic conditions than has been done so far. This thesis extends the knowledge about action recognition processes into more realistic scenarios and the far visual periphery. In four studies, life size action stimuli were used (I) to examine the action categorization abilities of central and peripheral vision, (II) to investigate the viewpoint-dependency of peripheral action representations, (III) to behaviorally measure the perceptive field sizes of action sensitive channels and (IV) to investigate the influence of additional actors in the visual scene on action recognition processes. The main results of the different studies can be summarized as follows. In Study I a high categorization performance for social actions throughout the visual field with a nonlinear performance decline towards the visual periphery was shown. Study II revealed a viewpoint–dependence of action recognition only in far visual periphery. In Study III large perceptive fields for action recognition were measured that decrease in size towards the periphery. And in Study IV no influence of a surrounding crowd of people on the recognition of actions in central vision and the visual periphery was shown. In sum, this thesis provides evidence that the abilities of peripheral vision have been underestimated and that peripheral vision might play a more important role in daily life than merely triggering gaze saccades to events in our environment.

# ACKNOWLEDGEMENTS

I spent the past three and a half years to discover the scientific world as well as my own capabilities. This thesis is the result. My experience at the Max-Planck-Institute for Biological Cybernetics in Tübingen has been nothing short of amazing. I am indebted to a number of people that made this time of my life highly memorable and marvelous.

First and foremost I wish to thank my supervisors Dr. Stephan de la Rosa and Dr. Isabelle Bülthoff, who did an incredible job in guiding me through this time and helped me with my first steps into the scientific world. I am aware that many people believe to have had the best supervisors in the world but I am certain that mine indeed are better than all the others! One simply couldn't wish for friendlier or better supervisors. How much I have learned from them, both scientifically and personally, is beyond words. Stephan has supported me throughout my thesis with his patience and immense knowledge whilst allowing me the room to work in my own way. He has been a constant source of encouragement, enthusiasm and inspiration. He helped shaping not only this work but also my future and my personality. Isabelle's support and her inspiring suggestions have been precious for the development of this thesis. Without her critical questions and her literary sensitiveness I would have never learned how to express my thoughts and none of my work would be written in a way that other people can understand it. She has been a truly dedicated mentor who always provided insightful discussions about research and creative comments to my work.

I would like to express my gratitude towards Prof. Dr. Heinrich Bülthoff, who managed to create a working atmosphere at the institute that allows people to thrive and enjoy science. The possibilities that he offers (for example the state-of-the-art technical equipment amongst many other things) and also his moral support helped me to become an enthusiastic scientist. I am grateful for the opportunity to be a part of his team.

Furthermore I would like to thank Prof. Dr. Martin Giese and Dr. Andreas Bartels, who were part of my advisory board, for steering this work in the right direction with their critical comments and fruitful discussions (that usually opened up many more research questions).

Dr. Tobias Meilinger, who was my second group leader but felt more like a third supervisor, was always able to provide support, helpful discussions and create an amazing working environment. His skill of playing the 'Devil's Advocate' will help to improve many more research projects to come. I would also like to acknowledge Nick Barraclough for his contribution to my projects with innumerous skype meetings and discussions of our thoughts and ideas during conferences. This collaboration was a great experience.

A special thank you goes to Joachim Tesch, without whose help and patient explanations the Panolab and me would never have become friends.

I would like to thank all of my colleagues from the RecCat group and the SSC group, the current ones as well as the former ones. Both groups have been a source of friendship and good advice over the years. Aurelie Saulton inspired me with her attitude, her serenity and also with her energy and passion. Dr. Markus Leyrer, my 'Reutlingen connection', was a huge source of moral support and ensured that next to all the research, exercise was not forgotten. Dr. Dong-Seon Chang, the show master, provided loads of encouragement and has many talents I could profit and learn from. Marianne Strickrodt was always able to provide recreation with her amazing humor and her constant advice during all those wonderful evenings we spent together. Dr. Katharina Dobs was my office mate for quite some time and made the working (and also the time we didn't really work) very pleasant. She showed me that determination is the most important skill to reach a goal. Dr. Kathrin Kaulard was my office mate at the very beginning of my time at the institute. She directly made me feel welcome and helped me to overcome the time where I was simply overwhelmed with the scientific work. Thomas Hinterecker only recently became part of the team and already is able to provide a lot of good critics and lots of ideas. With all these colleagues, the fun and the laughing was put forward and work didn't feel like work but rather like recreation together with friends.

And last but definitely not least I am very grateful towards many people that are not part of the scientific community I have worked with. My mother Susann Fademrecht who has supported me throughout my life and loves me unconditionally. Her faith in me carries me even through rough times. My brother Felix Fademrecht is my tower of strength and the person I have admired throughout my life. His support and encouragement guided me through my work. A good support system is essential for surviving and enjoying the time as a PhD student. I am lucky to have friends who shared the thrill with me and cheered me through my PhD work. Dr. Silvia Fademrecht has not only become my sister-in-law but also a close friend. With her I was able to compare PhD student experiences and together we found many great ways to take our minds off work and science. Nadja Decker, Dorothee Leyener and Evdoxia Vargiami have been my friends for a long time and turned out to be the best moral and emotional support one could hope for.

I will remember this time as the time where I learned the most, discovered the most and evolved the most. And I hope the times to come will be just as amazing.

# TABLE OF CONTENTS

# 1 INTRODUCTION

## 1.1 Human actions

### 1.1.1 Why investigate human actions

Aristotle once said 'Man is by nature a social animal'. Indeed, the most dominant force that influences our thoughts, behavior, physiology and neural activity is the need to participate in our social environment. Our survival critically depends on other humans and their behavior. Especially infants are highly dependent on the care of others and require their help to learn about the world around them. Independently of direct survival need, we like to be surrounded by friends with whom we can share our lives and experiences. Obviously, evolutionary pressure has favored complex social behavior in humans and therefore the development of suitable brain structures to process and interpret it. As social beings we constantly engage in social interaction with other people. Social interactions can be verbal, for example in a conversational setting, or they can be non-verbal, like eye-contact between two people or the mere observation of another person's body movements. Immensely important for the perception of the social environment are the human visual system and the auditory system. The human visual system processes the information that is contained in visible light and interprets the received signals. This way usually a clear and detailed image of the surrounding is produced that enables the interaction with the environment. There are different types of social cues that need to be interpreted for successful social interactions; one of them is body motion. Interpreting the body movements of other people is a way to collect information about their intentions and inner states (Blake & Shiffrar, 2007; Hugill, Fink, & Neave, 2010; Troje, 2003) and allows the interaction with other people. For example, in social interactions we need to be able to perceive and recognize another person's actions in order to be able to react appropriately.

In the current thesis, the recognition of bodily motion was investigated, more precisely social actions as revealed by body motion. In the following I will point out reasons as to why the recognition of human actions is an important research topic. I will review some of the existing literature on action understanding and introduce why I investigated more specifically action recognition in peripheral vision in this work.

Visually recognizing the actions of another person is important for humans to successfully navigate through their social and physical environment (de la Rosa, Mieskes, Bülthoff, & Curio, 2013; Giese & Poggio, 2003; Giese, 2014; Rizzolatti, Fogassi, & Gallese, 2001). The ability to understand the intentions of others, expressed by their body movements, is fundamental for human social behavior. Therefore, humans are highly sensitive to the social cues that are conveyed by the movements of other humans (Becchio et al., 2012; Becchio, Sartori, Bulgheroni, & Castiello, 2008; Johansson, von Hofsten, & Jansson, 1980; Marina Pavlova, Krägeloh-Mann, Birbaumer, & Sokolov, 2002; Pollick, Paterson, Bruderlin, & Sanford, 2001; Runeson & Frykholm, 1983; Sartori, Becchio, & Castiello, 2011; Troje, 2003). Lacking this ability is proposed to be associated with socially isolating mental illnesses, for instance autism (Iacoboni, Molnar-Szakacs, Gallese, Buccino, & Mazziotta, 2005).

We are interested in the actions that are social and communicative and whose recognition might be relevant for social interactions. The recognition of someone's actions is important for any human to be able to react appropriately. Whether a person approaching has good intentions or bad ones, whether he or she wants to greet or to perform an attack, in all, the identification of human social actions is crucial for wellbeing and survival.

For the perception of another person's actions, the human visual system is the most important sensory input. When it comes to the behavior of others, visual information can give us insights into both their active and inner states. In this context, the recognition of human actions is crucial for our social functioning. Understanding the recognition of actions is an important research area crossing the borders between scientific disciplines from computer vision to neuroscience (Jhuang & Serre, 2007). Humans need to be able to recognize actions under varying conditions, as for example changes in luminance, contrast, size, position, viewpoint and actor identity. Hence, the representation of human actions in the brain must be coded in such an invariant manner that accurate and reliable identification of the behavior of other people is possible within complex and changing social environments (Barraclough & Jellema, 2011).

## 1.1.2 How are actions recognized

The ability to understand another person's actions and to infer their intentions is important for humans to successfully navigate through their social and physical environment (Blakemore, Decety, & Albert, 2001; Call & Tomasello, 2008; C. D. Frith & Frith, 2005; C. Frith, 2002; U. Frith & Frith, 2003; Gallagher & Frith, 2015; Kilner & Frith, 2008). Many accounts provide ideas and insights about the processes underlying action recognition. For example, cognitive psychology and philosophy try to understand the human cognitive mechanisms of action recognition, whereas cognitive neuroscience focuses on the investigation of brain mechanisms underlying action perception. Last, the fields of robotics or computational and mathematical neuroscience, involve research on modeling movements or perceptual abilities (Gentsch, Weber, Synofzik, Vosgerau, & Schütz-Bosbach, 2016). Despite the information gained by these different fields of research, functional mechanisms and neural circuits proposed for action understanding remain controversial (de Lange, Spronk, Willems, Toni, & Bekkering, 2008; Hickok, 2009; Kilner & Frith, 2008; Rizzolatti et al., 2001; Saxe, 2005).

There are mainly three alternative accounts for the understanding of actions and intentions. According to the first account, called *theory-theory,* people understand the mind of others by applying a commonsense theory of the mind, as there is no direct access to the mental states of others. With this basic theory of the minds of others, people are supposed to be able to explain the behavior of others, their desires and beliefs and are able to explain their decisions (Gallese & Goldman, 1998; Gopnik, 1993; Przyrembel, Smallwood, Pauen, & Singer, 2012).

Second, the *simulation theory* or *direct-matching hypothesis* (Gallese & Goldman, 1998; Goldman, 1992; Iacoboni et al., 2005; Oberman & Ramachandran, 2007; Rizzolatti et al., 2001). Simulation theory supposes that actions of others are understood through a direct-matching mechanism that matches an observed action to the motor representation of this action using the mirror neuron system. Mirror neurons are a class of neurons that are activated both when executing a specific action and when observing the same action performed by someone else (Gallese, Fadiga, Fogassi, & Rizzolatti, 1996; Rizzolatti et al., 2001; Rizzolatti, 2005). In the brain, the rostral part of the inferior parietal lobule and the lower part of the precentral gyrus as well as the posterior part of the inferior frontal gyrus (IFG) are the core regions of the human mirror-neuron system (Rizzolatti & Craighero, 2004). If action recognition would be based purely on analysis of visual input, action understanding would mainly be mediated by the activity of the extrastriate visual areas, the inferotemporal lobe and the superior temporal sulcus (STS; Rizzolatti et al., 2001). The fact that motor areas are activated as well during action observation might lead to the conclusion that the motor system is highly involved in the understanding of another person's actions (Casile, Caggiano, & Ferrari, 2011; Iacoboni et al., 2005; Keysers & Gazzola, 2010; Oztop, Kawato, & Arbib, 2013; Rizzolatti & Craighero, 2004; Rizzolatti, 2005; Sinigaglia, 2013). Kilner and colleagues (2007) proposed an integration of the mirror neuron system within a predictive coding network in the sense that actions and their intentions are recognized by minimizing the prediction error along the different levels of cortical hierarchy, via reciprocal connections between the different cortical levels. This predictive coding framework considers a specific role of the mirror neuron system in understanding human actions and their intentions and therefore might explain how humans can infer another person's intentions by observing their actions.

An alternative account often described as *mentalizing* (the ability to ascribe mental states like intentions, beliefs and desires to oneself and to others, also referred to as *theory of mind*) led to the *visual hypothesis* (Brass, Schmitt, Spengler, & Gergely, 2007; F. P. de Lange et al., 2008; C. D. Frith & Frith, 2006; Liepelt, Von Cramon, & Brass, 2008; Rizzolatti et al., 2001; Van Overwalle & Baetens, 2009). The visual hypothesis describes action understanding as a result of pure visual analysis of the main elements a human action consists of (e.g. form, motion). Computational models provide a fundamental understanding of the underlying processes of visual action recognition. Giese et al. (2003) for example postulated a bottom-up controlled approach that is divided in a dorsal and a ventral processing stream and which uses learned prototypical patterns for the recognition of human actions. These patterns can be considered to be snapshot sequences of either body form (ventral form pathway) or of complex optic flow (dorsal motion pathway). According to this action recognition model, both processing streams contain hierarchies of so called neural feature detectors which process form features in the ventral stream and optic flow features in the dorsal stream. The complexity of these features increases along the hierarchy. This model explains the fast recognition of human actions in a feed-forward hierarchy without the need of top-down influences. This model has been mainly verified with locomotive actions (i.e. walking). Fleischer and colleagues (2013) on the other hand move further than locomotive actions and consider goal-directed hand movements. They developed a computational model of visual processes that are at play when recognizing actions directed towards an object (e.g. reach-to-grasp movements). Another computational model in line with the visual hypothesis is provided by Lange and Lappe (2006). Here it is assumed that static template cells of human walkers are used for a template-matching approach. The processing hierarchy in this model is divided in two stages, a static form stage followed by a dynamic form stage. At the static form stage, only form information is analyzed in each frame. In this stage, the temporal order of the frames is neglected. In the dynamic form stage, the global motion is processed and the frame order is analyzed. By this integration of dynamic form information over time, an action sequence is being recognized. In comparison to the approaches of Giese and Poggio (2003) as well as Fleischer and colleagues (2013), where the processing of local motion signals is being considered, for the approach of Lange and Lappe (2006) local motion signals are not critical for the recognition of biological motion, since template matching can be achieved by global form analysis. Later on, Theusner and colleagues (2014) proposed a model that uses a combination of posture-selective neurons (encoding specific postures) and neurons selective for body motion (encoding bodily action through the sequence of body postures, based on standard motion detectors). This model is able to show that standard mechanisms like spatio-temporal filters which are considered as part of motion detection mechanisms, can be applied in a novel manner to acquire a high-level analysis of human body movements (Theusner et al., 2014).

Whether the activation of motor areas is a requirement for action recognition or not, and whether visual and motor processes have the same importance for recognizing human actions, has not been answered so far. Casile and Giese (2006) provide evidence that motor learning of an unknown action, without visual input, directly influences the visual recognition performance of the same action. de la Rosa and colleagues (2016) investigated the interaction of visual and motor information for the understanding of human actions, when they observe and execute actions simultaneously, as it usually often happens in social interaction scenarios in real life. Their results show that when a person observes an action and executes an action at the same time, the recognition of actions mainly relies on visual mechanisms and is not influenced by their motor representations. Only in passive viewing conditions (i.e. observing without executing an action), contribution of the motor system to action recognition was found. The interaction of visual system and motor system in the recognition of human actions is not yet completely understood though and requires further investigation.

Research on visual action perception has mainly concentrated on visual perception in central vision. However, in daily life situations, actions most often appear in a person's visual periphery. Investigating action recognition processes in central vision as well as in the visual periphery might increase the understanding of action recognition mechanisms that are at play in real life scenarios.

## 1.2 Peripheral vision

### 1.2.1 Why investigate the visual periphery?

The human visual field covers between 200° and 220° of visual angle horizontally and 150° vertically (Harrington, 1981). Foveal vision, where visual information is received with the highest resolution, amounts to only 2° (in diameter) of the visual field (Strasburger, Rentschler, & Jüttner, 2011). The remaining visual field (99%) is considered as peripheral vision, where visual information is perceived with progressively lower resolution with increasing eccentricity. Nevertheless, the neural processes underlying peripheral vision - especially the visual abilities of the far periphery - are not well understood and have largely been neglected in visual research.

Input from the visual periphery is of high importance in daily-life situations as can be inferred from the many problems people suffering from tunnel vision struggle with. For example, these people have difficulties in crossing the streets, because they are rarely able to judge a gap in traffic as being big enough for them to cross the street and are therefore more insecure than people with intact peripheral vision (Cheong et al. 2008). Peripheral vision also plays a role in the execution of reaching and grasping movements. At the beginning of these movements, humans usually fixate the object they are trying to reach while their reaching arm is visible in the visual periphery. Visual input about arm's position has been shown to be quite important for accuracy of these movements (Sivak & MacKenzie, 1990). Moreover, occluding peripheral vision during locomotion leads, among other effects, to a decrease in walking speed (Graci, Elliott, & Buckley, 2009). A loss of the peripheral visual field affects the representation of space as well. For example, errors in objects' placement and the estimation of an object's location in space, increase with decreasing visual field (Fortenbaugh, Hicks, Hao, & Turano, 2007). In addition, input from the visual periphery is believed to have a large impact on the perception of the emotional content of visual stimuli, especially the emotional content of visual stimuli (e.g. fearful faces) in peripheral vision influences perception. Multiple studies indeed found evidence of an unconscious or implicit processing for emotional stimuli in peripheral vision (Bayle, Henaff, & Krolak-Salmon, 2009; Rigoulot et al., 2011; Rigoulot, D'Hondt, Honoré, & Sequeira, 2012). Threat-related stimuli, for example fearful faces, are quickly detected and are processed by a subcortical route involving the magnocellular system (Bayle et al. 2009), which is essentially afferented by the peripheral retina. All these studies indicate the importance of the visual periphery for perception and action in daily life, despite the low resolution of the peripheral retina.

## 1.2.2 Physiological differences in central and peripheral vision

The fovea is a small part of the retina (about 1.5 mm in diameter) that is characterized by a higher amount of cone cells and a higher density of ganglion cells compared to the visual periphery (Shapiro, Knight, & Lu, 2011). The decline of visual acuity from the fovea towards the periphery is well investigated (for comprehensive reviews, see Kerr, 1971; M Millodot, 1972). Visual resolution drops off drastically with increasing distance from the fovea (Larson & Loschky, 2009). The reason for this degradation in visual acuity with increasing eccentricity has frequently been ascribed to neural factors such as increased receptive field size and decline in cone density (Michel Millodot, Johnson, Lamont, & Leibowitz, 1975). The density of photoreceptors in the retina (especially cone cells) is much higher in central vision compared to peripheral vision. Additionally, far greater pooling of information from the individual photoreceptors by retinal ganglion cells occurs in the visual periphery than in central vision, leading to a decrease of visual resolution in the periphery. Many more cells in the lateral geniculate nucleus (LGN) and the primary visual cortex (V1) represent central vision compared to peripheral vision. This is called cortical magnification (Cowey & Rolls, 1974; Daniel & Whitteridge, 1961; Rovamo, Virsu, Laurinen, & Hyvärinen, 1982). Small visual stimuli can be better processed in central vision, whereas the same visual information must be magnified for perception in the visual periphery to obtain similar recognition performance (Rovamo & Virsu, 1979). In consequence people usually use foveal vision when trying to recognize persons and objects (Larson & Loschky, 2009).

## 1.2.3 Receptive fields vs perceptual fields

Since the increasing receptive field sizes towards the visual periphery have been associated with loss in visual acuity, understanding action recognition in peripheral vision might be partly explained by receptive field size. Neurons recorded in the STS of monkeys are reported to have large receptive fields that extent about 25° into both visual hemifields and typically include the fovea (Bruce, Desimone, Gross, & Gross, 1981; Perrett et al., 1989). However, little is known about receptive field sizes of action sensitive units in humans. Although physiological measures of receptive field sizes of action sensitive neural units would be an intuitive choice, they are hard to obtain in humans. As an alternative, spatial sampling areas of action sensitive mechanisms can be estimated using behavioral experiments (i.e. action adaptation paradigm).

The term *receptive field* of a neuron denotes its area of the visual field in which a visual stimulus will influence the response pattern of a neural unit. The receptive field of a ganglion cell in the retina consists of the input from the photoreceptors that build synapses with it. Going further, a group of ganglion cells then denotes the receptive field of the neuron they are connected with in the visual cortex. Towards the visual periphery, the average size of the receptive fields increase (Hubel & Wiesel, 1965; Wilson & Sherman, 1976). Receptive field sizes scale not only with increasing eccentricity, but scale also along the hierarchy of the visual pathways in the brain. Hence, neurons in higher processing areas in the brain pool information from multiple cells of the lower processing stages (Freeman & Simoncelli, 2011; Giese & Poggio, 2003) and their receptive size increases accordingly.

A *perceptive field* is considered to be the psychophysical correlate of the physiologically determined receptive field (Neri & Levi, 2006; Spillmann, 2014; Troscianko, 1982). Perceptive fields are behaviorally measurable as opposed to receptive fields for which physiological measurements are required to determine their size. The term was first introduced by Jung and Spillmann (1970), linking neural functioning to perceptual properties. It is assumed that perceptive fields are the summation of neural properties of all stages in the processing hierarchy of the brain that an observer uses to perform a certain visual task. Therefore, perceptive fields may be similar to one physiological receptive field or to the summation of many, depending on the task requirements. For example, for low level visual stimuli, like oriented bars or color stimuli, perceptive fields might be quite similar to receptive fields (e.g. in size), whereas for high-level stimuli the perceptive field might consist of the summation of multiple receptive fields. Hence, the link between the concept of a perceptive field and receptive fields may or may not be trivial, since perception is a complex process, involving different cortical areas (Neri & Levi, 2006). In the case of action perception, perceptive fields most certainly include rather a population of receptive fields than a single receptive field. In other words, receptive fields measure the spatial sampling area of a single neuron, whereas perceptive fields are the spatial sampling area for a population of neurons. Receptive and perceptive fields are therefore not identical but might correlate with each other.

## 1.2.4 Actions in the visual periphery

In a crowded area, other people appear in our visual periphery. Accordingly, actions are not only performed at our point of fixation but can appear somewhere in the visual field. Therefore, perception of human actions in the visual periphery is immensely important for our social behavior as it allows us to respond to action events that happen outside the central visual field.

Research concerning action perception in the visual periphery is limited to a few studies. Thompson and colleagues (2007) showed that direction discrimination of biological motion is possible in the visual periphery at 10° eccentricity. Nonetheless, when the biological motion stimuli were embedded in dynamic visual noise, stimulus detection performance was lower in peripheral vision compared to central vision. This indicates that peripheral vision suffers from a deficit in segregating signal from surrounding noise. Ikeda and colleagues (2005) showed that for biological motion stimuli, detection performance in noise could not be equated between central vision and peripheral vision (up to 12° eccentricity) by scaling of the stimulus size. Further, they showed that the inversion effect (i.e. higher performance for upright biological motion stimuli as compared to inverted stimuli) was not found in peripheral vision, but only in central vision. They conclude, that processing mechanisms for biological motion stimuli are focused on central vision. Gurnsey and colleagues (2008) on the other hand were able to equate participants' recognition performance of biological motion stimuli across the visual field by scaling the stimulus size. Even at 16° eccentricity a similar performance as in the central vision was achieved with a stimulus size of 20° visual angle, in a direction discrimination and a walker identification task. Ikeda and colleagues (2013) investigated in a different study the discrimination of walking direction of point light walkers surrounded by two flanking walkers at 5° eccentricity. With decreasing flanker target distance the discrimination of the walking direction of the central walker became increasingly difficult due to crowding effects. However, with scrambled flankers the crowding effect disappeared, suggesting that in the case of biological motion stimuli crowding effects occur at a high-level of motion information processing. In the study of de Lussanet and colleagues (2008) participants reported the facing direction of a biological motion stimulus. Walkers facing to the right were better recognized in the right visual hemifield (up to 20° eccentricity) whereas walkers facing to the left were superior in the left visual hemifield. Since motor and somatosensory cortical areas usually represent the contralateral side of the body and visual areas get input from the contralateral hemifield as well, they concluded that the hemispheric specialization of a person's own body map also represents the bodies of other people.

The commonality of these studies is that only the near visual periphery has been examined. In order to understand the contribution of the visual periphery to visual perception larger eccentricities need to be investigated as well. Many questions arise when we consider the far visual periphery. Are humans able to recognize actions in their far visual periphery or is this part of the retina devoted simply to the detection of events and initiating gaze saccades to the periphery? What aspects of an action can be perceived in the periphery? Can people only perceive the emotional gist of an action or the broad category the action belongs to or is the visual information provided by the peripheral retina, sufficient enough to identify the action in the periphery? Since motion information is one of the key components of human actions, in the following, amongst other things, it will be argued that the dynamic nature of human actions might be of special importance for action recognition in the visual periphery.

## 1.2.5 Important factors that might influence action recognition in the periphery under realistic viewing conditions

### 1.2.5.1 Motion perception in the visual periphery

Previous research provides evidence for a difference in motion processing mechanisms for central and peripheral vision (Cormack, Blake, & Hiris, 1992; Lewis, Rosén, Unsbo, & Gustafsson, 2011; Traschütz, Zinke, & Wegener, 2012). For example, motion perception in the periphery is tuned to lower spatial frequencies and higher speeds compared to central vision (Johnston & Wright, 1985; Koenderink & van Doorn, 1978; McKee & Nakayama, 1984; Virsu, Rovamo, & Laurinen, 1982). Similarly, single cell recordings in macaque cortical areas V1 and V2 also reveal that cells with foveal receptive fields prefer slower speeds and those in the periphery prefer faster speeds (Orban, Kennedy, & Bullier, 1986). In contrast to those studies, Lappin and colleagues (2009) reported greater similarities of foveal and peripheral motion sensitivities as they found that for speeds above 0.5 deg/s, motion detection thresholds were not correlated with physiological factors that limit acuity in fovea and periphery.

By definition, human actions are dynamic and evolve over time. At some point during their actions, the actor may reach a pose that allows the identification of the action independent of any motion information. We describe this pose as *key-frame* (de la Rosa et al., 2013). Indeed, Lange and Lappe (2006) propose in their model of biological motion perception that global form information is critical and sufficient for the correct recognition of a walker. They state that the presence of local motion does not enhance performance in a direction discrimination task. The model assumes a recognition of static action snapshots that are integrated over time. Their assumptions could indicate that in case of human social actions, static images of an action are recognizable when they show a posture that allows identification of the action (key-frame). The ability to use key-frames to recognize human actions has been confirmed behaviorally, physiologically, and was part of algorithms for computer vision (Carlsson & Sullivan, 2001; Coulson, 2004; Jellema & Perrett, 2003; Laptev & Pérez, 2007; Sullivan & Carlsson, 2002). Nonetheless, whether action recognition mechanisms rely mainly on form or motion information or require both equally, is still highly controversial. The view that action recognition mechanisms rely mainly on form information (Beintema, Georg, & Lappe, 2006; Beintema & Lappe, 2002; Bertenthal & Pinto, 1994; J. Lange, Georg, & Lappe, 2006; J. Lange & Lappe, 2006) is challenged by the view that biological motion perception is mainly driven by integration of motion signals (Casile & Giese, 2005; Fleischer et al., 2013; Giese & Poggio, 2003; Neri, Luu, & Levi, 2006; Theusner et al., 2014).

However, is the presence of motion information helpful for the recognition of human actions in the visual periphery? It has not yet been investigated whether motion information is the key to recognizing actions or whether key-frames are sufficient for action recognition in the visual periphery. This is one of the questions that I will investigate in this thesis.

## 1.2.5.2 Crowding in the visual periphery

The term crowding refers to a deleterious effect on recognition of visual targets due to the presence of other objects next to the one to recognize, presumably caused by the decline of visual acuity towards the periphery (Levi, 2008; Pelli & Tillman, 2008; Whitney

& Levi, 2011). Crowding limits visual perception and recognition throughout most of the visual field outside of the fovea and impairs the ability to recognize and respond appropriately to objects and actions in clutter (Ikeda et al., 2013; Whitney & Levi, 2011). When participants are able to recognize a single small letter in their visual periphery, they have difficulty to recognize that letter when it is flanked by other letters. As previously mentioned, receptive field sizes of ganglion cells increase towards the visual periphery and two stimuli that are processed by the same receptive field are more difficult to dissociate from each other. This could be one explanation for crowding effects and could also explain why crowding occurs mainly in peripheral vision. Critical for the occurrence of crowding effects is usually the critical distance between the target and the flankers. Crowding happens when the target-flanker separation is smaller than the critical distance. Bouma (1970) empirically determined the critical separation distance for which crowding is reliably observed as half the eccentricity of the stimulus' presentation location and with that formed a rule of thumb that has proven to be quite reliable (Chung, Levi, & Legge, 2001; Kooi, Toet, Tripathy, & Levi, 1994; Toet & Levi, 1992).

Despite a great deal of research, the mechanisms underlying crowding are not yet fully understood (Levi, 2008; Pelli & Tillman, 2008; Whitney & Levi, 2011). Flom and colleagues stated in their early work (Flom, Weymouth, & Kahneman, 1963) that the distance over which spatial interaction occurs depends on the size of the receptive fields that are involved in the recognition of the target. In the visual periphery, larger receptive fields are found compared to central vision and this scale shift might result in larger crowding distances. Crowding is different from other seemingly similar phenomena, for example masking, lateral interaction and surround suppression. These phenomena are distinct from each other, possibly relying on different neural processes (Levi, 2008).

In action recognition, crowding has been studied in the context of biological motion with point-light walkers (Ikeda et al., 2013; Thornton & Vuong, 2004). However, little is known about the effect of crowding on the human ability to distinguish different actions.

## 1.3 Aim and structure of the thesis

In the following, I will highlight the aim of the current thesis and explain the main purposes for the four studies that are part of the thesis.

Although the literature provides already a lot of information to understand the recognition process for human actions, the investigation of human actions usually takes place with rather simplified or reduced stimuli. These stimuli range from pictures or videos of a single movement or even of a single body part (i.e. the hand) to biological motion stimuli that consist mostly of motion information only. With these kind of stimuli researchers gained deep insights into the recognition process of human actions. Investigated were the recognition of the actor's identity (Cutting & Kozlowski, 1977; Loula, Prasad, Harber, & Shiffrar, 2005), intention (Runeson & Frykholm, 1983) or sex (Barclay, Cutting, & Kozlowski, 1978; Kozlowski & Cutting, 1977). This research already shows quite impressively that humans can use another person's body movement to make various judgments about this person. When including form information people often reduce their stimuli as well to a level where mostly only one body part is seen, for example in the investigation of hand actions (Barraclough, Keith, Xiao, Oram, & Perrett, 2009; Fleischer et al., 2013; Iacoboni et al., 2005).

Differences between desktop and real life visual conditions invite the question how action recognition processes perform under real life conditions. The advantage of desktop experiments is that researchers have full control over the experimental manipulations and therefore can directly relate the observed effect to the experimental conditions. Although these studies give important insights into the recognition of human actions and lay the groundwork for understanding the underlying processes, researching action recognition with highly reduced stimuli is very far away from the perception of human actions in real life. The quality of an experiment is denoted by its capacity to demonstrate cause-effect relationships. Therefore, the experimental conditions must eliminate all other possible causes, which often results in overly artificial situations far away from real world experiences and hence lacking generalizability and ecological validity. In order to understand the mechanisms underlying human action recognition in real life, one must consider the conditions under which action recognition usually occurs. When recognizing the potential shortcomings of laboratory experiments with widely reduced stimuli, experimental designs can be developed that enhance the usefulness of lab-generated data. In order to step into this direction, several aspects need to be taken into account like for example, the fact that actions might occur in any area of the visual field of view, the fact that actions need to be categorized in real life instead of being only detected and that actions can be seen with different orientations.

This thesis consists of three studies that investigate different aspects that might play a role in human action recognition in the visual periphery. The aim of this dissertation was to advance our understanding of action recognition processes throughout the visual field under close to natural conditions. In order to move towards a higher ecological validity, we investigated action recognition in central vision and the visual periphery under more naturalistic conditions, by using life sized dynamic action stimuli that were recorded from different actors and therefore contain the natural variation in movement styles of different persons. Beginning with an exploratory study, I investigated the action recognition abilities of the far visual periphery. Followed by a closer characterization of the representation of peripherally presented actions in the brain, examining the viewpoint sensitivity of action recognition processes throughout the visual field. In order to gain a deeper understanding of the underlying neural processes that enable the recognition of actions in the far visual periphery, in the third study I measured the perceptive field size of action sensitive neural channels with a behavioral paradigm (i.e. action adaptation paradigm). After having gained knowledge about the perceptive field sizes in central and peripheral vision, in the fourth study I examined the relationship between crowding and perceptive field size by presenting a crowd of people in the visual field that surrounded the target actor, as it is often the case in real life situations.

The following part of the current thesis introduces and summarizes the different studies. In the general discussion the results of the studies are discussed with respect to the most relevant literature.

## 1.3.1 Study I

The aim of Study I was to gain first insights into the abilities of the visual periphery concerning the recognition and categorization of human actions. In Experiment 1, we asked the question: What can we recognize about an action in the visual periphery? Experiment 2 gives answers to the question: Does the recognition of actions depend on the analysis of the kinematic content of the action? We investigated these four questions using different recognition tasks and measuring the recognition performance of social actions throughout the horizontal visual field using natural size human-like avatars.

### 1.3.1.1 Categorization and recognition levels

Previous literature on action recognition mostly involved tasks that are not directly related to the recognition of the action. Tasks that have been frequently used are for example direction discrimination tasks (Barraclough & Jellema, 2011; Bertenthal & Pinto, 1994; Gurnsey, Roddy, & Troje, 2010; Gurnsey & Troje, 2010; J. Lange & Lappe, 2006; Oram & Perrett, 1996; B. Thompson et al., 2007; Thurman & Lu, 2013) or action detection in noise (Ikeda et al., 2005; Manera, Becchio, Schouten, Bara, & Verfaillie, 2011; Neri et al., 2006; Maria Pavlova & Sokolov, 2000; B. Thompson et al., 2007). However, in situations of social interaction, it is quite important to correctly categorize the other person's action, which rests on the ability to tell different actions apart. For example, only if a person is able to distinguish a high five action from a hitting action can this person perform an appropriate response and either respond to the high five or protect him- of herself from being hurt. Therefore, investigating action categorization has more implications for real life conditions than action detection or direction discrimination tasks. However, categorization is a rather complex process that has been investigated mainly for objects. The recognition of an object occurs on different levels. Take for example a table. Here the word table marks the basic level. The object can also be described as furniture, which denotes a more general (superordinate) level or it can be described at a more detailed level as a desk for example, which indicates the subordinate level of recognition (Rosch, Mervis, Gray, Johnson, & Boyes-braem, 1976). In this line, it has also been shown that these three levels of recognition require different amounts of visual information for the categorization process. Judging an object as being a desk (subordinate level), for example, requires more detailed information about the object than categorizing it to be a table (basic level; Rosch et al., 1976; Tanaka & Curran,

2001). Therefore, the basic level is considered as the entry level, the word table comes first in mind, before we recognize an object as a desk or as furniture.

Similar to objects, human actions can be recognized at different levels as well. A handshake action can for example be described as a greeting or, on a more detailed level, one could say it is a handshake (de la Rosa, Choudhery, et al., 2014). For the description of these recognition levels, we refer to the specificity of the actions. This means that for the more specific group of actions (e.g. handshake) we use the term 'first level' and for the more general group of actions (e.g. greeting) we use the term 'second level'. Similar to the categorization of objects, the different action recognition levels are associated with different recognition performance.

A handshake action can be described at a more abstract level as a positive action. Whether the emotional valence indeed denotes a superordinate level of action recognition is still an open question. However, there is no doubt in the importance of the emotional valence of an action. Compared to neutral stimuli, emotional stimuli are faster detected and drive more attentional resources. Due to their relevance for human social behavior, emotional stimuli are expected to be detected quickly. The literature on emotion processing provides evidence of fast emotional processing of visual stimuli even if they are presented very briefly or even subliminally (Calvo & Esteves, 2005; Dimberg, Thunberg, & Elmehed, 2000; Hermans, Spruyt, De Houwer, & Eelen, 2003; Öhman & Soares, 1998). The differentiation between neutral and unpleasant stimuli occurs early in visual processing as event-related potential (ERP) studies have shown (70–120 ms after stimulus onset; see (Keil et al., 2001; Pourtois, Thut, De Peralta, Michel, & Vuilleumier, 2005). These findings suggest a preferential processing of emotional stimuli even when the emotional meaning of the stimulus is task irrelevant (Fox, Russo, Bowles, & Dutton, 2001). Therefore, even if the emotional valence is not the equivalent of the superordinate level in action recognition, this quality can still be used to recognize different actions. The categorization in positive and negative actions might be already helpful enough for initiating an adequate reaction, for example, in case of a negative action, one can assume a threat and take precautions to avoid suffering. A correct identification of an action in the visual periphery would enable promptly planning the appropriate action response. Therefore, in order to investigate the contribution of peripheral vision to human social interactions in Study I the action recognition abilities of the peripheral visual field are examined.

### 1.3.1.2 Motion energy as a cue

Considering the difference in processing mechanisms for motion perception in central and peripheral vision the question arises whether the motion information contained in human actions naturally is an important key feature that is used for recognition. For example, negative actions are usually executed fast. Therefore, these actions contain mostly a higher amount of motion energy than most actions with positive valence. In order to investigate whether motion energy is the main component of the actions that participants use to categorize actions, the recognition performance for dynamic actions needs to be compared to the recognition of static actions. In Experiment 2 of Study I, the influence of motion information on action recognition processes in central and peripheral vision was investigated by comparing the recognition of dynamic action stimuli with recognition performance for static images.

In this study, I investigated participants' action recognition performance throughout the visual field in three categorization tasks. In a second experiment the role of motion information was assessed by comparison of action recognition performance for dynamic and static action stimuli.

## 1.3.2 Study II

Study II to characterize the underlying neural representations of social actions in the brain. I examined one property that is commonly found for the perception of visual stimuli, namely viewpoint dependency. Previous research shows that actions are represented in a viewpoint-dependent manner in the brain (Daems & Verfaillie, 1999; de la Rosa et al., 2013; Gurnsey et al., 2010; Jokisch, Daum, & Troje, 2006; Perrett et al., 1989; Verfaillie, 1993). However, it has been suggested that the integration of visual action information in the fovea differs from that in the periphery (Thompson et al., 2007; Thurman & Lu, 2013). Therefore, we investigated the viewpoint-dependency of action perception in central and peripheral vision.

### 1.3.2.1 Viewpoint dependency of object and action representation

When we see other people performing actions, we are confronted with the actions from many different viewpoints. Human actions are inherently three dimensional and can therefore provide an infinite number of different views and due to their dynamic nature they also provide different appearances along the time line. Although action recognition needs to be very precise in order to correctly discriminate between different actions, their representation needs to be robust enough to allow recognition from many different viewpoints.

The discussion whether the representation of objects, faces and actions is view-dependent or not is still ongoing. On the one hand Biederman and Gerhardstein (1993) postulate a structural-description theory according to which visual recognition is viewpoint-invariant within a certain range of viewpoints provided that all of these views show the same major component parts (geons) of the object and their qualitative spatial relations. On the other hand, the image-based theory, introduced by Bülthoff and Edelman (1993; Bülthoff & Edelman, 1992; Edelman & Bülthoff, 1992), argues for a viewpoint-dependent recognition of visual objects. This theory suggests a viewpoint-dependent mechanism and encourages a multiple-view approach in the sense that objects might be encoded as a set of view-specific representations, matched to percepts using mental rotation or normalization procedures to transform the visual image to the closest known view. In order to explain results from psychophysical experiments, neurophysiology studies and computer vision, aspects of both theories have been combined (Foster & Gilson, 2002; Tarr & Bülthoff, 1998). Many objects in daily life as well as faces and human actions are usually seen from many different viewpoints and people are well trained in their recognition. There is strong evidence that the recognition of human actions is indeed viewpoint-dependent. Participants show a viewpoint-independent performance in recognizing one's own walking patterns whereas for other individuals recognition performance was higher for frontal view, compared to half-profile and profile view presentations (Gurnsey et al., 2010; Jokisch et al., 2006). Priming stimuli seen from the same viewpoint as the test stimuli are more effective than mirror-image priming stimuli (Daems & Verfaillie, 1999). Verfaillie (1993) examined the effects of depth rotation using short-term priming with point-light walkers and were able to show that priming effects only occurred when the priming walker and the test walker had the same orientation. Viewpoint-dependent recognition performance was also found for social actions occurring between pairs of people (e.g. shaking hands; de la Rosa et al. 2013). Physiological evidence for viewpoint-dependent recognition of human bodies has been provided by Perrett and colleagues (1989), who used single-cell recordings to localize cells in the temporal cortex that are only activated when seeing faces or bodies from a particular viewpoint. These results indicate that the recognition of human actions is viewpoint-dependent even though human actions are hardly unfamiliar stimuli. In addition to these results, Caggiano and colleagues (2015) measured local field potentials in monkey area F5, while the animal was presented with goal-directed actions either from a first-person perspective (i.e. as if carrying out the action themselves) or a third-person perspective (i.e. seeing someone else performing the action). They found significant differences between first- and third-person perspectives with a superiority of the first-person view. However, one cannot assume that visual mechanisms identified in the fovea also apply to the visual periphery. Hence viewpoint dependency could be different for foveal and peripheral vision.

**1.3.2.2 Viewpoint dependency due to a feeling of engagement in social interaction**
Different viewpoints could lead to different recognition performance because of the social relevance of the stimulus. Viewing an action directed toward us (front view) might facilitate our impression of being the recipient of the action (first-person perspective), whereas when the action is seen as directed in another direction (e.g. profile view) it might give a feeling of being a detached observer of an action that is directed somewhere else (third-person perspective). This assumption rests upon the postulate that social perception is fundamentally different in situations where we are part of an interaction between people and situations where we represent an impartial observer (Schilbach et al., 2013). However, it remains mostly unclear whether neural processes are manipulated by the degree to which a person feels involved in social interaction and also whether the neural networks involved complement each other or are completely disconnected (Schilbach, 2010). Behavioral and neuroimaging results of Schilbach and colleagues (2006) show that participants are biased towards giving socially relevant facial expressions a higher rating when they were directed towards the observer and

that neural activation show different patterns when the facial are directed towards the participants than when they are directed elsewhere. Similarly, actions that are directed towards an observer might be more salient than actions directed away from the observer. This increase in saliency of an action when viewed frontally, may play an even more important role for action recognition processes in the visual periphery.

In this study, participants' action recognition performance was examined for six different actions that were presented either in the front view or the profile view at different positions in the visual field.

## 1.3.3 Study III

Study III examined the neural basis of action recognition in terms of perceptual field size. Here, I used an action adaptation paradigm to selectively target action sensitive perceptual channels and measure the spatial extent of the sampling area of action sensitive processes (perceptive fields) at different positions of the visual field.

Like receptive fields, perceptive fields become larger with increasing eccentricity (Ransom-Hogg & Spillmann, 1980; Spillmann, 2014). A receptive field consists of a central disk, the receptive field center, and the surround, a concentric ring region around the center. Results from experiments with monkeys show that perceptive field centers (measured psychophysically) are about the same size as receptive field centers, (measured with physiological methods; Spillmann, 2014). It is assumed that the relationship between receptive fields and perceptive fields in humans is similar to that in monkeys (Spillmann, Ransom-Hogg, & Oehler, 1987). Measuring perceptive field sizes of action sensitive units gives insights into the perceptual processes that underlie the recognition performance for social actions appearing in foveal and peripheral vision. To that end, we made use of a well-established psychophysical tool, the adaptation aftereffect, to measure perceptual fields.

### 1.3.3.1 Adaptation aftereffects

Exposing observers to a visual stimulus for a prolonged amount of time (adaptation) can transiently change the subsequent percept of an ambiguous test stimulus. This effect is quickly evident in the case of color adaptation for example. After adapting to a green square, a white square is perceived with a reddish tint (adaptation aftereffect). Adaptation effects occur when the visual processes between adaptor and test stimulus are shared (e.g. the pooled response across several color channels). Adaptation is believed to alter the tuning characteristics of visual processes involved in the perception of the adaptor. If these processes are partially shared between adaptor and test stimulus, these alterations are passed on to the perception of the test stimulus thereby changing its percept (Webster 2011). Systematic variation of the visual resemblance between adaptor and test stimulus allows adaptation aftereffects to be used to assess the tuning characteristics of visual processes. This method has therefore also been called the psychophysicist's microelectrode (Frisby, 1979). Early work on visual adaptation aftereffects focused on low-level stimulus properties such as color, motion, and orientation (C.W.G. Clifford, 2002; Gibson & Radner, 1937; M. A. Webster & Leonard, 2008). However, in recent decades scientists have started to explore high-level properties in terms of adaptation. Most of this research has focused on face perception. For example, reliable adaptation aftereffects have been demonstrated for the perception of facial characteristics, such as sex, attractiveness, ethnicity, and identity (Leopold, O'Toole, Vetter, & Blanz, 2001; Rhodes, Jeffery, Watson, Clifford, & Nakayama, 2003; Rhodes, Lie, Ewing, Evangelista, & Tanaka, 2010; M. A. Webster, Kaping, Mizokami, & Duhamel, 2004). These studies suggest that adaptation is not a unique mechanism of the low-level sensory cortex, but can also target higher-level cortical areas and therefore makes the investigation of tuning characteristics of high-level recognition processes possible. Accordingly, adaptation paradigms have also been used for examination of visual processes underlying action perception. Previous research in this direction mainly focused on investigating the visual mechanisms regarding the perception of gender (Jordan, Fallah, & Stoner, 2006; Troje, Sadr, Geyer, & Nakayama, 2006) and emotions (Roether, Omlor, Christensen, & Giese, 2009) from biological motion, walking direction discrimination (Barraclough & Jellema, 2011) and weight judgments with object-directed actions (Barraclough et al., 2009). Considering the fact that action categorization is a highly essential process in social behavior that is needed to identify observed actions correctly and performing an appropriate response, it is peculiar that this high-level process has received much less attention. In fact, the visual mechanisms underlying action categorization are poorly understood. de la Rosa and colleagues (2014) conducted the first study that uses the adaptation methodology for the investigation of high-level influences on categorical action perception. Using an action adaptation paradigm they were able to show that action recognition mechanisms are modulated by social context. Participants categorized static images of ambiguous actions that were rendered from a video showing the body posture transition between a wave and a slap. By first showing a video scene with either friendly content (a person waving) or a scene with hostile content (a person slapping another person), the authors found that action adaptation aftereffects were modulated by social action context (friendly or hostile) that preceded the action although the physical properties of adaptor and test stimuli were unchanged. Their results suggest that action representations are defined by preceding events, respectively actions, supporting the idea that action categorization is modulated by high-level influences.

In Study III, the action adaptation paradigm was applied such that it allowed the measurement of perceptive field sizes for action recognition. In theory, observing an adaptation effect at a location in the visual field that is slightly different from the location where the adaptor was presented, would indicate that both locations belong to the same perceptive field of the neural population (also called action channel) that was adapted to the action. If an adaptation effect at another location than the adapted location is not present, we can deduct that this location lies outside of the perceptive field of the specific action channel. In this study participants were adapted to an action at one location and presented with the test stimulus at different locations in the visual field to measure the spatial extent of the adaptation effects and with this the spatial extent of perceptive fields of action sensitive neural channels.

## 1.3.4 Study IV

Study IV investigated whether action recognition processes in central and peripheral vision are influenced by the presence of other people in the visual scene. As previously mentioned, one explanation for crowding effects is that two stimuli that are processed by the same receptive field are more difficult to dissociate from each other. A visual scene, cluttered with multiple actors thus could influence the perception of the actions of a target actor.

In real life situations, acting humans are usually not alone but we see them often surrounded by other people performing different actions. Therefore, stepping towards a higher ecological validity requires examining the potential interference caused by the presence of other actors on the identification of the action of a target actor. In Study IV we investigated the influence of a crowd of people on action recognition processes using an action adaptation paradigm and an action recognition task. Adding in the scene individuals that perform various actions while standing close to the actor, could induce the well-known crowding effect, especially in the visual periphery. The deleterious effect on visual recognition of objects and actions due to a cluttered surrounding is described as crowding and is believed to be caused by the well-established decline of visual acuity towards the periphery (Levi, 2008). The effect of crowding on the perception of biological motion has been studied using point-light walkers. Using a direction discrimination task, Ikeda et al. (Ikeda et al., 2013) showed that crowding occurred only with walking flankers but not with scrambled ones. This indicates that crowding of biological motion is a high-level effect. In the experiment of Thornton and Vuong (2004), where participants were asked to discriminate the walking direction of the central walker while ignoring the flankers, they found that biological motion can be processed passively in a bottom-up fashion and therefore the flankers' walking direction influenced the perception of the target stimuli's walking direction. Investigating the influence of crowding on the human ability to discriminate different actions denotes a next step in the investigation of crowding effects on daily life perception. Applying clutter in the form of additional actors in the visual scene allows to investigate the degree to which visual clutter in the scene negatively impacts the visual processes underlying the ability to tell different actions apart. In this study, the influence of a crowd of people on participants' action perception was investigated using an action adaptation paradigm and a recognition task.

## 1.4 General discussion

The aim of this study was to investigate the recognition of social actions throughout the visual field under more naturalistic conditions than has been done before. To move towards a more realistic testing environment, I used a virtual reality setup that allowed the presentation of moving life-size human figures that were presented anywhere over the entire horizontal visual field. The use of life-size stimuli enabled the investigation of properties of visual action recognition processes close to real life situations. When increasing the realism of experimental conditions, there is usually a trade-off between maintaining a highly controllable setup and a less controllable realistic environment. Some researchers switch to field experiments, because they argue that this might be a good way of increasing the ecological validity of their studies. In field experiments, researchers investigate participants' behavior outside of the laboratory, in a natural setting. The participants in a field experiment are sometimes even unaware that they are in fact part of an experiment. Some researchers argue that the external validity of such an experiment is high because it is taking place in the real world. However, as real-world settings differ dramatically from each other, findings in one real-world setting may or may not generalize to another real-world setting. Field studies sometimes lack internal validity due to the fact that there are usually many factors that cannot really be controlled. Laboratory-generated data on the other hand allow to make strong conclusions from the acquired data but are quite artificial and lack ecological validity. In order to bridge the gap, I used a virtual reality setup. Virtual reality allows to move towards more realistic conditions without losing the advantages of a completely controllable experimental setup and therefore increases the generalizability of experimental results to real life situations.

Study I of this thesis was conducted to gain first insights into human action recognition abilities in far visual periphery. Because of the decreasing visual resolution with eccentricity, the visual periphery was mainly believed to be important for triggering gaze saccades towards suspicious events in our visual field. The results of Study I prove the contrary. Here, a surprisingly high recognition performance was found for social actions, even up to 60° eccentricity. Moreover, participants did not only perceive partial aspects of the action but received enough visual information to identify actions that were presented in their visual periphery. Most astonishingly the recognition performance did not decrease compared to central vision up to 30° eccentricity but built a plateau before decreasing with higher eccentricities. This indicates that the action recognition abilities in peripheral vision are very similar to foveal performance for a wide range of eccentricities. The relationship of recognition performance with eccentricity turned out to be nonlinear. This nonlinearity stands in stark contrast to findings for object recognition at such far visual eccentricities (Jebara, Pins, Despretz, & Boucart, 2009; Thorpe, Gegenfurtner, Fabre-Thorpe, & Bülthoff, 2001), where a linear decline of recognition performance with eccentricity was found. I suggest that these differences between my results and the previous research on object recognition in the visual periphery might result from motion information contained in our action stimuli, as the objects were presented as static pictures in the studies of Thorpe and colleagues (2001) and Jebara et al. (2009). This idea was investigated with Experiment 2 of Study I. Here, I compared the recognition of dynamic actions with the recognition of static images of the key-frames of the actions. The results showed a nonlinear relationship between recognition performance and eccentricity for both dynamic and static actions. This finding leads to the speculation that the underlying processes involved in action recognition might differ from object recognition processes.

Study I provided first insights into action recognition abilities of the far visual periphery, leading to the question arose whether the underlying processes of action recognition are different for central and peripheral vision. Previous research has already suggested that the integration of visual action information in the fovea differs from that in the periphery (Thompson et al., 2007; Thurman & Lu, 2013). For a deeper understanding of the non-linear action recognition performance in the periphery I examined the neural underpinnings of action representations in the brain. These aspects are described in the following.

In Study II, the viewpoint-dependency of action recognition processes was investigated. Actions were presented to the participants either in the front view or in the side (profile) view, ensuring different perspectives of the actions. The results showed shorter reaction times for actions seen side on, in far periphery from 30° eccentricity on. In central vision the action recognition performance was viewpoint independent. Research in central vision provides evidence that action recognition processes are viewpoint-dependent (Daems & Verfaillie, 1999; de la Rosa et al., 2013; Gurnsey et al., 2010; Jokisch et al., 2006; Verfaillie, 1993). However, the experimental conditions of the aforementioned studies were far less realistic than the stimuli and setup used in my experiments. One could argue that any differences between the results might be owed to the amount of realism in the provided viewing conditions, for example the stimulus size.

The side view of the stimuli we used in the study provided more visible motion information. This could explain the viewpoint-dependency of the visual periphery in terms of a need for more visual information, as provided by the side view of the tested actions. Therefore, despite the high action recognition performance of the peripheral visual system, compared to the fovea, peripheral action recognition processes benefit from additional visual information, which could explain the differences found for the different viewpoints.

The processing of visual information is strictly tied to the physiological properties of the visual system (e.g. receptive field size, distribution of photoreceptors in the retina, cortical magnification). We know that the visual periphery is characterized by a decreasing number of cone photoreceptors, a decreasing density of ganglion cells and increasing receptive field sizes. Increasing receptive fields lead to a lower resolution in the periphery, which should decrease the ability to pick up details that are important for action discrimination. With regards to those facts, how can we explain the high action recognition performance that was found in Experiment 2 despite those reduced physiological properties in far periphery? In Study III, I measured behaviorally the perceptive field sizes of action sensitive perceptual channels. The action adaptation paradigm was applied as a useful tool to selectively target action sensitive channels and investigate their area of sensitivity in the visual field. The results revealed a large perceptive field in central vision (62.74° of visual angle) and decreasing perceptive fields towards the periphery (at -20° eccentricity: 29.06°; at -40° eccentricity: 25.72° visual angle) for action recognition processes. This finding is surprising in light of what is known about the change of receptive field sizes with eccentricity. However, considering the literature, the measured perceptive field sizes are in the same order of magnitude as previously reported receptive field sizes. For example, Oram & Perrett (1994) reported receptive fields of about 20° in the anterior STS for biological motion stimuli. Ito and colleagues (1995) found cells with receptive field sizes of about 25° in anterior inferior temporal cortex (IT) for pattern recognition. In V4 receptive field sizes are expected to extend between 5° and 10° at 10° eccentricity (Gattass, Sousa, & Gross, 1988). The perceptive field sizes resulting from Study III are much larger. However, a perceptive field might consist of multiple receptive fields and therefore might be appropriate. The large perceptive field in central vision overlaps with peripheral perceptive fields to a high degree. The high action recognition performance up to 30° eccentricity could therefore be attributed to the influence of central vision processes that also sample visual information of the peripheral retina. Therefore, recognition processes of central vision that influence peripheral viewing might overwrite certain differences between central and peripheral processing mechanisms. This concept could also explain the viewpoint-invariance up to 15° eccentricity that was shown in Study II. Recognition processes of central vision could influence the recognition up to 15° and decrease the need for additional visual information to recognize an action. In using the action adaptation paradigm to behaviorally measure the perceptive field sizes I tried to target neural units that are specifically prone to recognize a certain human action.

A flanker task is another paradigm that could be used to assess perceptive field sizes for action recognition processes. In a flanker task a target stimulus is flanked by two stimuli with high similarity to the target. When the distance between the target and the flankers is sufficiently small, the flankers interfere with the perception of the target. By varying the target flanker separation, the critical distance that leads to interference with the perception of the target can be determined. Similar paradigms, for example the Westheimer paradigm (Ransom-Hogg & Spillmann, 1980; Spillmann, 2014) or the Herman grid (Troscianko, 1982), have been used to behaviorally assess the receptive field sizes for low level stimuli. In theory, stimuli that are presented in the same spatial area of sensitivity of a neural channel (e.g. receptive field or perceptive field), should be processed together and integrated by the neural channel. As soon as stimuli are not presented in the area of sensitivity of a neural channel, they are processed by different neural channels and therefore should not interfere perceptually. Since the receptive fields increase in size towards the visual periphery, perceptual interference occurs already with larger distances between target and flanking stimuli. As mentioned above, perceptual interference between target and flanking stimuli due to a small target-flanker separation is a phenomenon called crowding. Crowding occurs for a number of stimuli and throughout the visual field, however, more pronounced in the visual periphery (Kooi et al., 1994; Levi & Carney, 2009; Levi, Hariharan, & Klein, 2002; Levi, 2008; Pelli & Tillman, 2008; Whitney & Levi, 2011). Researchers believe inappropriate integration of target and flankers to be the reason for crowding effects to occur (Levi & Carney, 2009). However, the underlying mechanisms of crowding effects are still unknown, as Levi concluded in a recent review (Levi, 2008): 'Crowding is an enigma wrapped in a paradox and shrouded in a conundrum. Despite a great deal of new (and old) work, we do not yet have a full understanding of crowding'. This statement raises the question whether a flanking task would indeed be suitable to assess the spatial extent of perceptual fields for social action recognition. In Study IV, the influence of additional actors in the visual scene on the action recognition processes has been investigated. The results showed no significant effect of a crowd on the adaptation aftereffects as well as on the recognition performance as measured by accuracy and reaction times. The distance between the target actor and the closest crowd members (flankers) was small enough for the action stimuli to overlap and should therefore have led to crowding effects. Especially when we consider the spatial extent of the perceptive field in central vision that was measured in Study III. Multiple actors would have appeared in the same perceptual field, increasing the possibility of the stimuli to interfere perceptually. A possible explanation for the absence of crowding effects even though the perceptive field is large and would have contained multiple actors might be a dynamically adjusted perceptive field size depending on the amount of actions that need to be discriminated from each other in the visual field. In addition to this, there was no significant influence of the flanking stimuli on the adaptation aftereffect. Thus, indicating that the adaptation paradigm and flanker tasks measure different and independent characteristics of action recognition processes.

Perceptive fields could be used to predict human recognition performance, as has been suggested by Neri and Levi (2006). In order to assess whether perceptive field sizes can be linked to recognition performance, in Study III, I developed a probably oversimplified model that shows a relationship between the perceptive field size and the recognition performance in Study I. Gaussian functions were used to mathematically visualize the perceptive field properties, analogue to the visualization of receptive fields, which are usually described with a difference-of-Gaussian function. The *Full Width at Half Maximum* (FWHM) was defined as spatial extent of the perceptive field. These Gaussian functions were then used to describe the relationship between perceptive field size and participants' recognition performance in Study I.

A summation of the Gaussian functions represented a good fit of the recognition performance with an $R^2$ of 0.99. This basic relationship between the recognition performance and the perceptive field size is able to predict action recognition performance at any given point in the visual field. However, the definite interpretation of the parameters is yet to be determined. Whether a flanker task would also provide a measure for perceptive field sizes, leading to a direct link between perceptive field sizes and action recognition performance is not yet clear. On the one hand, the results of Study IV lead to the assumption that an assessment of perceptive field sizes would be unfruitful. On the other hand, the flanker task might simply be more effective for another stimulus class.

The absence of a crowded perception in Study IV gives rise to speculations about the nature of peripheral visual processes. Crowding in the visual periphery is often associated with texture perception as a result of joint statistics computation of the input image (Balas, Nakano, & Rosenholtz, 2009; Rosenholtz, Huang, Raj, Balas, & Ilie, 2012). Although such a model might explain crowding effects when pattern recognition of simple patterns fails, one could speculate that such a model would not to be able to capture action recognition processes in the visual periphery. Actions are much more complex due to the inherent motion information. According to computational models (Fleischer et al., 2013; Giese & Poggio, 2003; J. Lange & Lappe, 2006; Theusner et al., 2014) action recognition relies on the perception of snapshot images of action postures that are then integrated over time. A texture percept of these action snippets would render the recognition of the multiple snapshots that an action contains too error prone to achieve a high recognition performance in a crowded environment as shown in Study IV.

A possible assumption concerning the lack of influence of surrounding people on action recognition performance in Study IV, could be that the stimuli presented to one perceptive field are not automatically processed together and integrated but are processed hierarchically. The stimulus that drives the most attentional resources (the stimulus participants are told to attend to) might be prioritized in the recognition process, whereas the surrounding disregarded stimuli are then processed with less priority. A principle also known as biased competition theory. In real life the visual field usually contains many different objects or many other people that all need to compete for neural processing. In order to reduce the workload to the available capacity, attentional mechanisms limit the processing to items that are currently relevant for the behavior. These attentional mechanisms enhance the responses of neurons representing stimuli that are most relevant and can have bottom-up influences (e.g. higher contrast, higher saliency) or top-down influences (i.e. selective attention; Desimone, 1998; Kastner & Ungerleider, 2001; Mather & Sutherland, 2011). The target stimulus in Study IV was presented in front of the crowd and participants were instructed to attend to the target stimulus. This could have activated both kinds of attentional mechanisms, bottom-up and top-down influences, enabling biased competition in favor of the target stimulus. Thus, decreasing the influence of flanking distractors on the recognition of the target.

## 1.5 Conclusions and future work

In my thesis I investigated action recognition in central and peripheral vision by presenting life-size dynamic action stimuli to various locations along the whole horizontal visual field of my participants. These more naturalistic presentation conditions of action stimuli led to results that are quite different from laboratory-generated data. I was able to show that action recognition in the visual periphery is surprisingly good and could reveal that some characteristics of action recognition processes might differ for social actions in real life compared to the results gained in laboratory experiments with the use of a computer screen setup. More specifically, the results suggest that peripheral vision plays a more important role in our daily social interactions than merely triggering gaze saccades to conspicuous events in our environment. This thesis gives first insights into high-level visual processes in the visual periphery and lays the ground work for future investigations. A further assessment of perceptive field properties throughout the visual field for action recognition would be necessary in order to achieve a deeper understanding of the visual processes that are at play in real life scenarios. Some of the results provided in this thesis lead to the assumption that for object recognition and action recognition the underlying processes might differ in certain ways. A direct comparison between object and action recognition under more naturalistic conditions would be necessary to tease the processes apart. An aspect that has not been examined in this thesis, is the role of attentional top-down control. To be able to spread one's attention over the whole field of view, as is the case for my participants in this work, might result in different findings than in scenarios where the attention would be fixed on something else in the visual field. In sum, this thesis provides a different understanding of action recognition processes as to what has been known about action perception throughout the visual field and might lead to an explanation as to why human observers are still far better at recognition tasks than any computer vision routine (Balas et al., 2009).

## 1.6 References

Amano, K., Wandell, B. a, & Dumoulin, S. O. (2009). Visual field maps, population receptive field sizes, and visual field coverage in the human MT+ complex. *Journal of Neurophysiology*, *102*(5), 2704–18. http://doi.org/10.1152/jn.00102.2009

Anderson, S. J., Mullent, K. T., Hesst, R. F., Anderson, S. J., Mullen, K. T., & Hess, R. F. (1991). Human Peripheral Spatial Resolution for Achromatic and Chromatic Stimuli: Limits Imposed By Optical and Retinal Factors. *Journal of Physiology*, *442*, 47–64.

Balas, B., Nakano, L., & Rosenholtz, R. (2009). A summary-statistic representation in peripheral vision explains visual crowding. *Journal of Vision*, *9*(12), 13.1–18. http://doi.org/10.1167/9.12.13

Banks, M. S., Sekuler, A. B., & Anderson, S. J. (1991). Peripheral spatial vision: limits imposed by optics, photoreceptors, and receptor pooling. *J. Opt. Soc. Am. A*, *8*(11), 1775–1787.

http://doi.org/10.1364/JOSAA.8.001775

Barclay, C. D., Cutting, J. E., & Kozlowski, L. T. (1978). Temporal and spatial factors in gait perception that influence gender recognition. *Perception & Psychophysics*, *23*(2), 145–152. http://doi.org/10.3758/BF03208295

Barraclough, N. E., & Jellema, T. (2011). Visual Aftereffects for Walking Actions Reveal Underlying Neural Mechanisms for Action Recognition. *Psychological Science*, *22*(1), 87–94. http://doi.org/10.1177/0956797610391910

Barraclough, N. E., Keith, R. H., Xiao, D., Oram, M. W., & Perrett, D. I. (2009). Visual adaptation to goal-directed hand actions. *Journal of Cognitive Neuroscience*, *21*(9), 1806–1820. http://doi.org/10.1162/jocn.2008.21145

Bayle, D. J., Henaff, M.-A., & Krolak-Salmon, P. (2009). Unconsciously perceived fear in peripheral vision alerts the limbic system: a MEG study. *PloS One*, *4*(12), e8207. http://doi.org/10.1371/journal.pone.0008207

Bayle, D. J., Schoendorff, B., Hénaff, M.-A., & Krolak-Salmon, P. (2011). Emotional facial expression detection in the peripheral visual field. *PloS One*, *6*(6), e21584. http://doi.org/10.1371/journal.pone.0021584

Becchio, C., Cavallo, A., Begliomini, C., Sartori, L., Feltrin, G., & Castiello, U. (2012). Social grasping: From mirroring to mentalizing. *NeuroImage*, *61*(1), 240–248. http://doi.org/10.1016/j.neuroimage.2012.03.013

Becchio, C., Sartori, L., Bulgheroni, M., & Castiello, U. (2008). Both your intention and mine are reflected in the kinematics of my reach-to-grasp movement. *Cognition*, *106*(2), 894–912. http://doi.org/10.1016/j.cognition.2007.05.004

Beintema, J. A., Georg, K., & Lappe, M. (2006). Perception of biological motion from limited-lifetime stimuli. *Perception & Psychophysics*, *68*(4), 613–24. http://doi.org/10.3758/BF03208763

Beintema, J. A., & Lappe, M. (2002). Perception of biological motion without local image motion. *Proceedings of the National Academy of Sciences of the United States of America*, *99*(8), 5661–3. http://doi.org/10.1073/pnas.082483699

Bertenthal, B. I., & Pinto, J. (1994). Global Processing of Biological Motions. *Psychological Science*, *5*(4), 221–224. http://doi.org/10.1111/j.1467-9280.1994.tb00504.x

Biederman, I., & Gerhardstein, P. C. (1993). Recognizing depth-rotated objects: evidence and conditions for three-dimensional viewpoint invariance. *Journal of Experimental Psychology. Human Perception and Performance*, *19*(6), 1162–1182. http://doi.org/10.1037/h0090355

Blake, R., & Shiffrar, M. (2007). Perception of human motion. *Annual Review of Psychology*, *58*, 47–73. http://doi.org/10.1146/annurev.psych.57.102904.190152

Blake, R., Tadin, D., Sobel, K. V, Raissian, T. a, & Chong, S. C. (2006). Strength of early visual adaptation depends on visual awareness. *Proceedings of the National Academy of Sciences of the United States of America*, *103*(12), 4783–8. http://doi.org/10.1073/pnas.0509634103

Blakemore, S., Decety, J., & Albert, C. (2001). From the Perception of Action to the Understanding of Intention. *Nature Reviews Neuroscience*, *2*(August), 561–567. http://doi.org/10.1038/35086023

Bouma, H. (1970). Interaction Effects in Parafoveal Letter Recognition. *Nature*, *226*.

Brass, M., Schmitt, R. M., Spengler, S., & Gergely, G. (2007). Investigating Action Understanding: Inferential Processes versus Action Simulation. *Current Biology*, *17*(24), 2117–2121. http://doi.org/10.1016/j.cub.2007.11.057

Bruce, C., Desimone, R., Gross, C. G., & Gross, G. (1981). Visual properties of neurons in a polysensory area in superior temporal sulcus of the macaque . Visual Properties of Neurons in a Polysensory Area in Superior Temporal Sulcus of the Macaque. *Journal of Neurophysiology*, *46*(2), 369–384.

Bülthoff, H. H., & Edelman, S. (1992). Psychophysical support for a two-dimensional view interpolation theory of object recognition. *Proceedings of the National Academy of Sciences*, *89*, 60–64.

Bülthoff, H. H., & Edelman, S. (1993). Evaluating object recognition theories by computer graphics psychophysics. *Exploring Brain Functions: Models in Neuroscience*, 139–164. Retrieved from http://citeseerx.ist.psu.edu/viewdoc/download?doi=10.1.1.45.2993&rep=rep1&type=pdf \nhttp://citeseerx.ist.psu.edu/viewdoc/summary?doi=10.1.1.45.2993

Bülthoff, I., Bülthoff, H., & Sinha, P. (1998). Top-down influences on stereoscopic depth-perception. *Nature Neuroscience*, *1*(3), 254–257. http://doi.org/10.1038/699

Caggiano, V., Giese, M., Thier, P., & Casile, A. (2015). Encoding of point of view during action observation in the local field potentials of macaque area F5. *European Journal of Neuroscience*, *41*(4), 466–476. http://doi.org/10.1111/ejn.12793

Call, J., & Tomasello, M. (2008). Does the chimpanzee have a theory of mind? 30 years later. *Trends in Cognitive Sciences*, *12*(5), 187–192. http://doi.org/10.1016/j.tics.2008.02.010

Calvo, M., & Esteves, F. (2005). Detection of emotional faces: low perceptual threshold and wide attentional span. *Visual Cognition*, *12*(1), 13–27. http://doi.org/10.1080/13506280444000094

Carlsson, S., & Sullivan, J. (2001). Action recognition by shape matching to key frames. *Workshop on Models versus Exemplars in Computer Vision*, *1*, 18. Retrieved from http://ftp1.nada.kth.se/pub/CVAP/reports/ws_mod_ex_sull_car.pdf

Casile, A., Caggiano, V., & Ferrari, P. F. (2011). The Mirror Neuron System: A Fresh View. *The Neuroscientist*, *17*(5), 524–538. http://doi.org/10.1177/1073858410392239

Casile, A., & Giese, M. A. (2005). Critical features for the recognition of biological motion. *Journal of Vision*, *5*(4), 348–360. http://doi.org/10.1167/5.4.6

Casile, A., & Giese, M. A. (2006). Nonvisual motor training influences biological motion perception. *Current Biology*, *16*(1), 69–74. http://doi.org/10.1016/j.cub.2005.10.071

Chung, S. T. L., Levi, D. M., & Legge, G. E. (2001). Spatial-frequency and contrast properties of crowding. *Vision Research*, *41*(14), 1833–1850. http://doi.org/10.1016/S0042-6989(01)00071-2

Clifford, C. W. G. (2002). Perceptual adaptation: motion parallels orientation. *Trends in Cognitive Sciences*, *6*(3), 136–143. http://doi.org/10.1016/S1364-6613(00)01856-8

Clifford, C. W. G., Wyatt, A. M., Arnold, D. H., Smith, S. T., & Wenderoth, P. (2001). Orthogonal adaptation improves orientation discrimination. *Vision Research*, *41*(2), 151–159. http://doi.org/10.1016/S0042-6989(00)00248-0

Cormack, R., Blake, R., & Hiris, E. (1992). Misdirected visual motion in the peripheral visual field. *Vision Research*, *32*(1), 73–80. Retrieved from http://www.ncbi.nlm.nih.gov/pubmed/1502813

Coulson, M. (2004). Attributing emotion to static body postures: reconigition accuracy, confusion and view point dependence. *Journal of Nonverbal Behavior*, *28*(2), 117–139. http://doi.org/10.1023/B:JONB.0000023655.25550.be

Cowey, a, & Rolls, E. T. (1974). Human cortical magnification factor and its relation to visual acuity. *Experimental Brain Research. Experimentelle Hirnforschung. Experimentation Cerebrale, 21*(5), 447–454. http://doi.org/10.1007/BF00237163

Cutting, J. E., & Kozlowski, L. T. (1977). Recognizing friends by their walk: Gait perception without familiarity cues. *Bulletin of the Psychonomic Society, 9*(5), 353–356.

Daems, A., & Verfaillie, K. (1999). Viewpoint-dependent Priming Effects in the Perception of Human Actions and Body Postures. *Visual Cognition, 6*(6), 665–693. http://doi.org/10.1080/135062899394894

Daniel, P. M., & Whitteridge, D. (1961). The representation of the visual field on the cerebral cortex in monkeys. *Journal of Physiology, 159*, 203–221.

de la Rosa, S., Choudhery, R. N., & Chatziastros, A. (2011). Visual object detection, categorization, and identification tasks are associated with different time courses and sensitivities. *Journal of Experimental Psychology. Human Perception and Performance, 37*(1), 38–47. http://doi.org/10.1037/a0020553

de la Rosa, S., Choudhery, R. N., Curio, C., Ullman, S., Assif, L., & Bülthoff, H. H. (2014). Visusal categorization of social interactions. *Visual Cognition, 22*(9-10), 1233–1271. http://doi.org/10.1080/13506285.2014.991368

de la Rosa, S., Ferstl, Y., & Bülthoff, H. H. (2016). Visual adaptation dominates bimodal visual-motor action adaptation. *Scientific Reports, 6*(October 2015), 23829. http://doi.org/10.1038/srep23829

de la Rosa, S., Mieskes, S., Bülthoff, H. H., & Curio, C. (2013). View dependencies in the visual recognition of social interactions. *Frontiers in Psychology, 4*(October), 752. http://doi.org/10.3389/fpsyg.2013.00752

de la Rosa, S., Streuber, S., Giese, M. A., Bülthoff, H. H., & Curio, C. (2014). Putting Actions in Context: Visual Action Adaptation Aftereffects Are Modulated by Social Contexts. *PLoS ONE, 9*(1), e86502. http://doi.org/10.1371/journal.pone.0086502

de Lange, F. P., Spronk, M., Willems, R. M., Toni, I., & Bekkering, H. (2008). Complementary Systems for Understanding Action Intentions. *Current Biology, 18*(6), 454–457. http://doi.org/10.1016/j.cub.2008.02.057

de Lussanet, M. H. E., Fadiga, L., Michels, L., Seitz, R. J., Kleiser, R., & Lappe, M. (2008). Interaction of visual hemifield and body view in biological motion perception. *European Journal of*

*Neuroscience*, *27*(2), 514–522. http://doi.org/10.1111/j.1460-9568.2007.06009.x

Desimone, R. (1998). Visual attention mediated by biased competition in extrastriate visual cortex. *Philosophical Transactions of the Royal Society of London. Series B, Biological Sciences*, *353*(1373), 1245–55. http://doi.org/10.1098/rstb.1998.0280

DiCarlo, J. J., & Cox, D. D. (2007). Untangling invariant object recognition. *Trends in Cognitive Sciences*, *11*(8), 333–341. http://doi.org/10.1016/j.tics.2007.06.010

Dimberg, U., Thunberg, M., & Elmehed, K. (2000). Unconscious facial reactions to emotional facial expressions. *Psychological Science : A Journal of the American Psychological Society / APS*, *11*(1), 86–89. http://doi.org/10.1111/1467-9280.00221

Dumoulin, S. O., & Wandell, B. a. (2008). Population receptive field estimates in human visual cortex. *NeuroImage*, *39*(2), 647–660. http://doi.org/10.1016/j.neuroimage.2007.09.034

Edelman, S., & Bülthoff, H. H. (1992). Orientation Dependence in the Recognition of Familiar and Novel Views of Three-Dimensional Objects. *Vision Research*, *32*(12), 2385–2400.

Ennis, F. a, & Johnson, C. a. (2002). Are high-pass resolution perimetry thresholds sampling limited or optically limited? *Optometry and Vision Science : Official Publication of the American Academy of Optometry*, *79*(8), 506–11. http://doi.org/00006324-200208000-00013 [pii]

Fademrecht, L., Bülthoff, I., & de la Rosa, S. (2016). Action recognition in the visual periphery. *Journal of Vision*, *16*(3), 1–14. http://doi.org/10.1167/16.3.33.doi

Fleischer, F., Caggiano, V., Thier, P., & Giese, M. A. (2013). Physiologically Inspired Model for the Visual Recognition of Transitive Hand Actions. *The Journal of Neuroscience*, *33*(15), 6563–6580. http://doi.org/10.1523/JNEUROSCI.4129-12.2013

Flom, M., Weymouth, F., & Kahneman, D. (1963). Visual Resolution and Contour Integration. *Journal of the Optical Society of America*, *53*(9), 1026–32. Retrieved from http://www.ncbi.nlm.nih.gov/pubmed/14065335

Fortenbaugh, F. C., Hicks, J. C., Hao, L., & Turano, K. a. (2007). Losing sight of the bigger picture: peripheral field loss compresses representations of space. *Vision Research*, *47*(19), 2506–20. http://doi.org/10.1016/j.visres.2007.06.012

Foster, D. H., & Gilson, S. J. (2002). Recognizing novel three-dimensional objects by summing signals from parts and views. *Proceedings. Biological Sciences / The Royal Society*, *269*(1503), 1939–47. http://doi.org/10.1098/rspb.2002.2119

Fox, E., Russo, R., Bowles, R., & Dutton, K. (2001). Do threatening stimuli draw or hold visual attention in subclinical anxiety? *Journal of Experimental Psychology. General*, *130*(4), 681–700. http://doi.org/10.1037/0096-3445.130.4.681

Freeman, J., & Simoncelli, E. P. (2011). Metamers of the ventral stream. *Nature Neuroscience*, *14*(9), 1195–1201. http://doi.org/10.1038/nn.2889

Frisby, J. P. (1979). Seeing. *Oxford University Press*. http://doi.org/10.1007/s007690000247

Frisen, L., & Glansholm, A. (1975). Optical and Neural Resolution in Peripheral Vision. *Investigative Ophtalmology*, *14*(7), 528–536.

Frith, C. (2002). Attention to action and awareness of other minds. *Consciousness and Cognition*, *11*(4), 481–487. http://doi.org/10.1016/S1053-8100(02)00022-3

Frith, C. D., & Frith, U. (2005). Quick guide Theory of mind. *Current Biology*, *15*(17), 644–645. http://doi.org/10.1016/j.cub.2005.08.041

Frith, C. D., & Frith, U. (2006). The Neural Basis of Mentalizing. *Neuron*, *50*(4), 531–534. http://doi.org/10.1016/j.neuron.2006.05.001

Frith, U., & Frith, C. D. (2003). Development and neurophysiology of mentalizing. *Philosophical Transactions of the Royal Society of London. Series B, Biological Sciences*, *358*(1431), 459–473. http://doi.org/10.1098/rstb.2002.1218

Gallagher, H. L., & Frith, C. D. (2015). Functional imaging of "theory of mind." *Trends in Cognitive Sciences*, *7*(2), 77–83. http://doi.org/10.1016/S1364-6613(02)00025-6

Gallese, V., Fadiga, L., Fogassi, L., & Rizzolatti, G. (1996). Action recognition in the premotor cortex. *Brain*, *119*(2), 593–609.

Gallese, V., & Goldman, A. (1998). Mirror neurons and the mind-reading. *Trens in Cognitive Sciences*, *2*(12), 493–501. http://doi.org/10.1016/S1364-6613(98)01262-5

Gardner, T. (1973). Evidence for Independent in Tachistoscopic Parallel Channels Perception The present work is also reported in Indiana Mathematical Psychological Pro- gram Report No . 72-4 , Indiana University , 1972 . *Cognitive Psychology*, *155*, 130–155.

Gattass, R., Sousa, a P., & Gross, C. G. (1988). Visuotopic organization and extent of V3 and V4 of the macaque. *The Journal of Neuroscience : The Official Journal of the Society for Neuroscience*, *8*(6), 1831–1845.

Gentsch, A., Weber, A., Synofzik, M., Vosgerau, G., & Schütz-Bosbach, S. (2016). Towards a common framework of grounded action cognition: Relating motor control, perception and cognition. *Cognition, 146*, 81–89. http://doi.org/10.1016/j.cognition.2015.09.010

Gibson, J. J., & Radner, M. (1937). Adaptation, after-effect and Contrast in the perception of tilted lines, 186–196.

Giese, M. A. (2013). Biological and Body Motion Perception. *The Oxford Handbook of Perceptual Organization*.

Giese, M. A. (2014). Skeleton model for the neurodynamics of visual action representations. *24th International Conference on Artificial Neural Networks, ICANN 2014, 8681 LNCS*, 707–714. http://doi.org/10.1007/978-3-319-11179-7-89

Giese, M. A., & Poggio, T. (2003). Neural mechanisms for the recognition of biological movements. *Nature Reviews. Neuroscience, 4*(3), 179–92. http://doi.org/10.1038/nrn1057

Goldman, A. I. (1992). In Defense of the Simulation Theory. *Mind & Language, 7*(1-2), 104–119. http://doi.org/10.1111/j.1468-0017.1992.tb00200.x

Gopnik, A. (1993). How we know our minds: The illusion of first-person knowledge of intentionality. *Behavioral and Brain Sciences, 16*(01), 1. http://doi.org/10.1017/S0140525X00028636

Graci, V., Elliott, D. B., & Buckley, J. G. (2009). Peripheral visual cues affect minimum-foot-clearance during overground locomotion. *Gait & Posture, 30*(3), 370–4. http://doi.org/10.1016/j.gaitpost.2009.06.011

Grezes, J. (1998). Top Down Effect of Strategy on the Perception of Human Biological Motion : A PET Investigation. *Cognitive Neuropsychology, 15*(December 2012), 553–582. http://doi.org/10.1080/026432998381023

Grill-Spector, K., & Kanwisher, N. (2005). Do you know what it is as soon as you know it is there? *Psychological Science, 16*(2), 152–160. http://doi.org/10.1167/8.6.512

Gurnsey, R., Roddy, G., Ouhnana, M., & Troje, N. F. (2008). Stimulus magnification equates identification and discrimination of biological motion across the visual field. *Vision Research, 48*(28), 2827–2834. http://doi.org/10.1016/j.visres.2008.09.016

Gurnsey, R., Roddy, G., & Troje, N. F. (2010). Limits of peripheral direction discrimination of point-light walkers. *Journal of Vision, 10*(15), 1–17. http://doi.org/10.1167/10.2.15.Introduction

Gurnsey, R., & Troje, N. F. (2010). Peripheral sensitivity to biological motion conveyed by first and second-order signals. *Vision Research*, *50*(2), 127–135. http://doi.org/10.1016/j.visres.2009.10.020

Hansen, T., Pracejus, L., & Gegenfurtner, K. R. (2009). Color perception in the intermediate periphery of the visual field. *Journal of Vision*, *9*(4), 26.1–12. http://doi.org/10.1167/9.4.26

Harrington, D. O. (1981). *The Visual Fields: A Textbook and Atlas of Clinical Perimetry*. St. Louis, MO: Mosby.

Hermans, D., Spruyt, A., De Houwer, J., & Eelen, P. (2003). Affective Priming With Subliminally Presented Pictures. *Canadian Journal of Experimental Psychology*, *57*(2), 97–114.

Hickok, G. (2009). Eight Problems for the Mirror Neuron Theory of Action: Understanding in Monkeys and Humans. *J Cogn Neurosci.*, *21*(7), 1229–1243. http://doi.org/10.1162/jocn.2009.21189.Eight

Hoffman, K. L., & Logothetis, N. K. (2009). Cortical mechanisms of sensory learning and object recognition. *Philosophical Transactions of the Royal Society B: Biological Sciences*, *364*(1515), 321–329. http://doi.org/10.1098/rstb.2008.0271

Hubel, D. H., & Wiesel, T. N. (1965). Receptive field and functional architechture in two nonstriate visual areas (18 and 19) of the cat. *Journal of Neurophysiology*, *28*, 229–289.

Hubel, D. H., & Wiesel, T. N. (1974). Uniformity of Monkey Striate Cortex: A Parallel Relationship between Field Size, Scatter, and Magnification Factor. *The Journal of Comparative Neurologyhe Journal of Comparative Neurology*.

Hudson, M., Nicholson, T., Simpson, W. a, Ellis, R., Bach, P., Hudson, M., … Ellis, R. (2015). One Step Ahead : The Perceived Kinematics of Others ' Actions Are Biased Toward Expected Goals. *Journal of Experimental Psychology: General*. Retrieved from http://dx.doi.org/10.1037/

Hugill, N., Fink, B., & Neave, N. (2010). The role of human body movements in mate selection. *Evolutionary Psychology : An International Journal of Evolutionary Approaches to Psychology and Behavior*, *8*(1), 66–89. Retrieved from http://www.ncbi.nlm.nih.gov/pubmed/22947780

Iacoboni, M., Molnar-Szakacs, I., Gallese, V., Buccino, G., & Mazziotta, J. C. (2005). Grasping the intentions of others with one's own mirror neuron system. *PLoS Biology*, *3*(3), 0529–0535. http://doi.org/10.1371/journal.pbio.0030079

Ikeda, H., Blake, R., & Watanabe, K. (2005). Eccentric perception of biological motion is unscalably poor. *Vision Research*, *45*(15), 1935–43. http://doi.org/10.1016/j.visres.2005.02.001

Ikeda, H., Watanabe, K., & Cavanagh, P. (2013). Crowding of biological motion stimuli. *Journal of Vision*, *13(4)*(20), 1–6. http://doi.org/10.1167/13.4.20.doi

Ito, M., Tamura, H., Fujita, I., & Tanaka, K. (1995). Size and Position Invariance of Neuronal Responses in Monkey Inferoterotemporal Cortex. *Journal of Neurophysiologyo*, *73*(1), 218–226.

Jacobs, R. J. (1979). Visual resolution and contour interaction in the fovea and periphery. *Vision Research*, *19*(11), 1187–1195. http://doi.org/10.1016/0042-6989(79)90183-4

Jebara, N., Pins, D., Despretz, P., & Boucart, M. (2009). Face or building superiority in peripheral vision reversed by task requirements. *Advances in Cognitive Psychology*, *5*, 42–53. http://doi.org/10.2478/v10053-008-0065-5

Jellema, T., & Perrett, D. I. (2003). Cells in monkey STS responsive to articulated body motions and consequent static posture: A case of implied motion? *Neuropsychologia*, *41*(13), 1728–1737. http://doi.org/10.1016/S0028-3932(03)00175-1

Jhuang, H., & Serre, T. (2007). A biologically inspired system for action recognition. *Computer Vision, 2007. ICCV 2007. IEEE 11th International Conference on*, 1–8. http://doi.org/10.1109/ICCV.2007.4408988

Johansson, G., von Hofsten, C., & Jansson, G. (1980). Event Perception. *Perception*, *31*, 27–63.

Johnston, A., & Wright, M. J. (1985). Lower Thresholds of Motion Gratings as a Function of Eccentricity and Contrast. *Vision Research*, *25*(2), 179–185.

Jokisch, D., Daum, I., & Troje, N. F. (2006). Self recognition versus recognition of others by biological motion: Viewpoint-dependent effects. *Perception*, *35*(7), 911–920. http://doi.org/10.1068/p5540

Jordan, H., Fallah, M., & Stoner, G. R. (2006). Adaptation of gender derived from biological motion. *Nature Neuroscience*, *9*(6), 738–739. http://doi.org/10.1038/nn1710

Jung, R., & Spillmann, L. (1970). Receptive-field estimation and perceptual integration in human vision. *Early Experience and Visual Information Processing in Perceptual and Reading Disorders*, 181–197.

Kastner, S., & Ungerleider, L. G. (2001). The neural basis of biased competition in human visual cortex. *Neuropsychologia*, *39*(12), 1263–1276. http://doi.org/10.1016/S0028-3932(01)00116-6

Keil, A., Müller, M. M., Gruber, T., Wienbruch, C., Stolarova, M., & Elbert, T. (2001). Effects of emotional arousal in the cerebral hemispheres: A study of oscillatory brain activity and event-related potentials. *Clinical Neurophysiology*, *112*(11), 2057–2068. http://doi.org/10.1016/S1388-2457(01)00654-X

Kerr, J. (1971). Visual resolution in the periphery. *Perception & Psychophysics*, *9*, 375–378. Retrieved from http://link.springer.com/article/10.3758/BF03212671

Keysers, C., & Gazzola, V. (2010). Social Neuroscience: Mirror Neurons Recorded in Humans. *Current Biology*, *20*(8), R353–R354. http://doi.org/10.1016/j.cub.2010.03.013

Kilner, J. M., Friston, K. J., & Frith, C. D. (2007). Predictive coding: an account of the mirror neuron system. *Cognitive Processing*, *8*(3), 159–166. http://doi.org/10.1007/s10339-007-0170-2

Kilner, J. M., & Frith, C. D. (2008). Action Observation: Inferring Intentions without Mirror Neurons. *Current Biology*, *18*(1), 32–33. http://doi.org/10.1016/j.cub.2007.10.063

Koenderink, J. J., & van Doorn, A. J. (1978). Visual detection of spatial contrast; influence of location in the visual field, target extent and illuminance level. *Biological Cybernetics*, *30*(3), 157–167. http://doi.org/10.1007/BF00337144

Kooi, F. L., Toet, A., Tripathy, S. P., & Levi, D. M. (1994). The effect of similarity and duration on spatial interaction in peripheral vision. *Spatial Vision*, *8*(2), 255–279. http://doi.org/10.1017/CBO9781107415324.004

Kovacs, G. (2005). Electrophysiological Correlates of Visual Adaptation to Faces and Body Parts in Humans. *Cerebral Cortex*, *16*(5), 742–753. http://doi.org/10.1093/cercor/bhj020

Kozlowski, L. T., & Cutting, J. E. (1977). Recognizing the sex of a walker from a dynamic point-light display. *Perception & Psychophysics*, *21*(6), 575–580.

Kravitz, D. J., Kriegeskorte, N., & Baker, C. I. (2010). High-level visual object representations are constrained by position. *Cerebral Cortex*, *20*(December), 2916–2925. http://doi.org/10.1093/cercor/bhq042

Lange, J., Georg, K., & Lappe, M. (2006). Visual perception of biological motion by form: A template-matching analysis. *Journal of Vision*, *6*, 836–849.

Lange, J., & Lappe, M. (2006). A model of biological motion perception from configural form cues. *The Journal of Neuroscience : The Official Journal of the Society for Neuroscience*, *26*(11), 2894–906. http://doi.org/10.1523/JNEUROSCI.4915-05.2006

Lappin, J. S., Tadin, D., Nyquist, J. B., & Corn, A. L. (2009). Spatial and temporal limits of motion perception across variations in speed, eccentricity, and low vision. *Journal of Vision*, *9*(30), 1–14. http://doi.org/10.1167/9.1.30.Introduction

Laptev, I., & Pérez, P. (2007). Retrieving actions in movies. *Proceedings of the IEEE International Conference on Computer Vision*. http://doi.org/10.1109/ICCV.2007.4409105

Larson, A. M., & Loschky, L. C. (2009). The contributions of central versus peripheral vision to scene gist recognition. *Journal of Vision*, *9*(10), 6.1–16. http://doi.org/10.1167/9.10.6

Leopold, D. A., O'Toole, A. J., Vetter, T., & Blanz, V. (2001). Prototype-referenced shape encoding revealed by high-level aftereffects. *Nature Neuroscience*, *4*(1), 89–94. http://doi.org/10.1038/82947

Levi, D. M. (2008). Crowding - An essential bottleneck for object recognition: A mini-review. *Vision Research*, *48*(5), 635–654. http://doi.org/10.1016/j.visres.2007.12.009

Levi, D. M., & Carney, T. (2009). Crowding in Peripheral Vision: Why Bigger Is Better. *Current Biology*, *19*(23), 1988–1993. http://doi.org/10.1016/j.cub.2009.09.056

Levi, D. M., Hariharan, S., & Klein, S. A. (2002). Suppressive and facilitatory spatial interactions in peripheral vision: peripheral crowding is neither size invariant nor simple contrast masking. *Journal of Vision*, *2*(2), 167–177. http://doi.org/10.1167/2.2.3

Lewis, P., Rosén, R., Unsbo, P., & Gustafsson, J. (2011). Resolution of static and dynamic stimuli in the peripheral visual field. *Vision Research*, *51*(16), 1829–34. http://doi.org/10.1016/j.visres.2011.06.011

Liepelt, R., Von Cramon, D. Y., & Brass, M. (2008). How do we infer others' goals from non-stereotypic actions? The outcome of context-sensitive inferential processing in right inferior parietal and posterior temporal cortex. *NeuroImage*, *43*(4), 784–792. http://doi.org/10.1016/j.neuroimage.2008.08.007

Loula, F., Prasad, S., Harber, K., & Shiffrar, M. (2005). Recognizing people from their movement. *Journal of Experimental Psychology. Human Perception and Performance*, *31*(1), 210–220. http://doi.org/10.1037/0096-1523.31.1.210

Manera, V., Becchio, C., Schouten, B., Bara, B. G., & Verfaillie, K. (2011). Communicative interactions improve visual detection of biological motion. *PLoS ONE*, *6*(1). http://doi.org/10.1371/journal.pone.0014594

Mather, G., Verstraten, F., & Anstis, S. M. (1998). *The motion aftereffect: A modern perspective*. Mit Press.

Mather, M., & Sutherland, M. R. (2011). Arousal-Biased Competition in Perception and Memory. *Perspectives on Psychological Science : A Journal of the Association for Psychological Science*, *6*(2), 114–33. http://doi.org/10.1177/1745691611400234

McCollough, C. (1965). Color Adaptation of Edge-Detectors in the Human Visual System. *Science*, *149*(3688), 1115–1116. http://doi.org/10.1126/science.149.3688.1115

McKee, S. P., & Nakayama, K. (1984). The Detection of Motion in the Peripheral Visual Field. *Vision Research*, *24*(1), 25–32.

Millodot, M. (1972). Variation of visual acuity in the central region of the retina. *The British Journal of Physiological Optics*, *27*(1), 24–28.

Millodot, M., Johnson, C. A., Lamont, A., & Leibowitz, H. W. (1975). Effect of dioptrics on peripheral visual acuity. *Vision Research*, *15*(12), 1357–1362. http://doi.org/10.1016/0042-6989(75)90190-X

Naïli, F., Despretz, P., & Boucart, M. (2006). Colour recognition at large visual eccentricities in normal observers and patients with low vision. *Neuroreport*, *17*(15), 1571–4. http://doi.org/10.1097/01.wnr.0000236864.54575.98

Nelson, W. W., & Loftus, G. R. (1980). The functional visual field during picture viewing. *Journal of Experimental Psychology. Human Learning and Memory*, *6*(4), 391–399. http://doi.org/10.1037/0278-7393.6.4.391

Neri, P., & Levi, D. M. (2006). Receptive versus perceptive fields from the reverse-correlation viewpoint. *Vision Research*, *46*(16), 2465–2474. http://doi.org/10.1016/j.visres.2006.02.002

Neri, P., Luu, J. Y., & Levi, D. M. (2006). Meaningful interactions can enhance visual discrimination of human agents. *Nature Neuroscience*, *9*(9), 1186–1192. http://doi.org/10.1038/nn1759

Oberman, L. M., & Ramachandran, V. S. (2007). The simulating social mind: the role of the mirror neuron system and simulation in the social and communicative deficits of autism spectrum disorders. *Psychological Bulletin*, *133*(2), 310–327. http://doi.org/10.1037/0033-

2909.133.2.310

Öhman, A., & Soares, J. J. F. (1998). Emotional Conditioning to Masked Stimuli: Expectancies for Aversive Outcomes Following Nonrecognized Fear-Relevant Stimuli. *Journal of Experimental Psychology: General*, *127*(1), 69–82. Retrieved from http://www.sciencedirect.com/science/article/B6X07-46NPTWC-5/2/2db8480244813fe4d5392266a525e2af

Op de Beeck, H., & Vogels, R. (2000). Spatial Sensitivity of Macaque Inferior Temporal Neurons. *The Journal of Comparative Neurologyhe Journal of Comparative Neurology*, *426*, 505–518.

Oram, M. W., & Perrett, D. I. (1994). Responses of Anterior Superior Temporal Polysensory (STPa) Neurons to "Biological Motion" Stimuli. *Journal of Cognitive Neuroscience*, *6*(2), 99–116. http://doi.org/10.1162/jocn.1994.6.2.99

Oram, M. W., & Perrett, D. I. (1996). Integration of form and motion in the anterior superior temporal polysensory area (STPa) of the macaque monkey. *Journal of Neurophysiology*, *76*(1), 109–129. Retrieved from http://www.ncbi.nlm.nih.gov/pubmed/8836213

Orban, G. A., Kennedy, H., & Bullier, J. (1986). Velocity sensitivity and direction selectivity of neurons in areas V1 and V2 of the monkey: influence of eccentricity. *Journal of Neurophysiology*, *56*(2), 462–480.

Oztop, E., Kawato, M., & Arbib, M. a. (2013). Mirror neurons: Functions, mechanisms and models. *Neuroscience Letters*, *540*, 43–55. http://doi.org/10.1016/j.neulet.2012.10.005

Pavlova, M., Krägeloh-Mann, I., Birbaumer, N., & Sokolov, A. (2002). Biological motion shown backwards: The apparent-facing effect. *Perception*, *31*(4), 435–443. http://doi.org/10.1068/p3262

Pavlova, M., & Sokolov, A. (2000). Orientation specificity in biological motion perception. *Perception & Psychophysics*, *62*(5), 889–899. http://doi.org/10.3758/BF03212075

Pelli, D. G., & Tillman, K. a. (2008). The uncrowded window of object recognition. *Nature Neuroscience*, *11*(10), 1129–1135. http://doi.org/10.1038/nn1208-1463b

Perrett, D. I., Harries, M. H., Bevan, R., Thomas, S., Benson, P. J., Mistlin, a J., … Ortega, J. E. (1989). Frameworks of analysis for the neural representation of animate objects and actions. *The Journal of Experimental Biology*, *146*, 87–113. Retrieved from http://www.ncbi.nlm.nih.gov/pubmed/2689570

Pollick, F. E., Paterson, H. M., Bruderlin, A., & Sanford, A. J. (2001). Perceiving affect from arm movement. *Cognition*, *82*(2), 51–61. http://doi.org/10.1016/S0010-0277(01)00147-0

Popovic, Z., & Sjöstrand, J. (2001). Resolution, separation of retinal ganglion cells, and cortical magnification in humans. *Vision Research*, *41*(10-11), 1313–1319. http://doi.org/10.1016/S0042-6989(00)00290-X

Popovic, Z., & Sjöstrand, J. (2005). The relation between resolution measurements and numbers of retinal ganglion cells in the same human subjects. *Vision Research*, *45*(17), 2331–2338. http://doi.org/10.1016/j.visres.2005.02.013

Pourtois, G., Thut, G., De Peralta, R. G., Michel, C., & Vuilleumier, P. (2005). Two electrophysiological stages of spatial orienting towards fearful faces: Early temporo-parietal activation preceding gain control in extrastriate visual cortex. *NeuroImage*, *26*(1), 149–163. http://doi.org/10.1016/j.neuroimage.2005.01.015

Przyrembel, M., Smallwood, J., Pauen, M., & Singer, T. (2012). Illuminating the dark matter of social neuroscience: Considering the problem of social interaction from philosophical, psychological, and neuroscientific perspectives. *Frontiers in Human Neuroscience*, *6*(June), 190. http://doi.org/10.3389/fnhum.2012.00190

Ransom-Hogg, A., & Spillmann, L. (1980). Perceptive field size in fovea and periphery of the light- and dark-adapted retina. *Vision Research*, *20*(3), 221–228. http://doi.org/10.1016/0042-6989(80)90106-6

Rhodes, G., Jeffery, L., Watson, T. L., Clifford, C. W. G., & Nakayama, K. (2003). Fitting the Mind to the World: Face Adaptation and Attractiveness Aftereffects. *Psychological Science*, *14*(6), 558–566. http://doi.org/10.1046/j.0956-7976.2003.psci_1465.x

Rhodes, G., Lie, H. C., Ewing, L., Evangelista, E., & Tanaka, J. W. (2010). Does perceived race affect discrimination and recognition of ambiguous-race faces? A test of the sociocognitive hypothesis. *Journal of Experimental Psychology: Learning, Memory, and Cognition*, *36*(1), 217–223. http://doi.org/10.1037/a0017680

Riesenhuber, M., & Poggio, T. a. (2000). Models of object recognition. *Nature Neuroscience*, *3*, 1199–1204. http://doi.org/10.1038/81479

Rigoulot, S., D'Hondt, F., Defoort-Dhellemmes, S., Despretz, P., Honoré, J., & Sequeira, H. (2011). Fearful faces impact in peripheral vision: behavioral and neural evidence. *Neuropsychologia*, *49*(7), 2013–21. http://doi.org/10.1016/j.neuropsychologia.2011.03.031

Rigoulot, S., D'Hondt, F., Honoré, J., & Sequeira, H. (2012). Implicit emotional processing in peripheral vision: Behavioral and neural evidence. *Neuropsychologia*, *50*(12), 2887–96. http://doi.org/10.1016/j.neuropsychologia.2012.08.015

Rizzolatti, G. (2005). The mirror neuron system and its function in humans. *Anatomy and Embryology*, *210*(5-6), 419–421. http://doi.org/10.1007/s00429-005-0039-z

Rizzolatti, G., & Craighero, L. (2004). the Mirror-Neuron System. *Annual Review of Neuroscience*, *27*(1), 169–192. http://doi.org/10.1146/annurev.neuro.27.070203.144230

Rizzolatti, G., Fogassi, L., & Gallese, V. (2001). Neurophysiological mechanisms underlying the understanding and imitation of action. *Nature Reviews Neuroscience*, *2*(September), 661–70. http://doi.org/10.1038/35090060

Roether, C. L., Omlor, L., Christensen, A., & Giese, M. A. (2009). Critical features for the perception of emotion from gait. *Journal of Vision*, *9*(6), 15.1–32. http://doi.org/10.1167/9.6.15

Rosch, E., Mervis, C. B., Gray, W. D., Johnson, D. M., & Boyes-braem, P. (1976). Basic Objects in Natural Categories. *Cognitive Psychology*, *8*, 382–439.

Rosenholtz, R., Huang, J., Raj, a., Balas, B. J., & Ilie, L. (2012). A summary statistic representation in peripheral vision explains visual search. *Journal of Vision*, *12*(4), 14–14. http://doi.org/10.1167/12.4.14

Rovamo, J., & Virsu, V. (1979). An estimation and application of the human cortical magnification factor. *Exp Brain Res*, *37*(3), 495–510.

Rovamo, J., Virsu, V., Laurinen, P., & Hyvärinen, L. (1982). Resolution of gratings oriented along and across meridians in peripheral vision. *Investigative Ophthalmology & Visual Science*, *23*(5), 666–70. Retrieved from http://www.ncbi.nlm.nih.gov/pubmed/7129811

Runeson, S., & Frykholm, G. (1983). Kinematic specification of dynamics as an informational basis for person-and-action perception: Expectation, gender recognition, and deceptive intention. *Journal of Experimental Psychology: General*, *112*(4), 585–615. http://doi.org/10.1037/0096-3445.112.4.585

Sartori, L., Becchio, C., & Castiello, U. (2011). Cues to intention: The role of movement information. *Cognition*, *119*(2), 242–252. http://doi.org/10.1016/j.cognition.2011.01.014

Saxe, R. (2005). Against simulation: The argument from error. *Trends in Cognitive Sciences*, *9*(4), 174–179. http://doi.org/10.1016/j.tics.2005.01.012

Sayres, R., Grill-spector, K., Hindy, N. C., Turk-browne, N. B., Witthoft, N., Nguyen, M. L., … Grill-spector, K. (2015). Relating Retinotopic and Object-Selective Responses in Human Lateral Occipital Cortex Relating Retinotopic and Object-Selective Responses in Human Lateral Occipital Cortex, (May 2008), 249–267. http://doi.org/10.1152/jn.01383.2007

Schilbach, L. (2010). A second-person approach to other minds. *Nature Reviews. Neuroscience*, *11*(6), 449. http://doi.org/10.1038/nrn2805-c1

Schilbach, L., Timmermans, B., Reddy, V., Costall, A., Bente, G., Schlicht, T., & Vogeley, K. (2013). Toward a second-person neuroscience. *The Behavioral and Brain Sciences*, *36*(4), 393–414. http://doi.org/10.1017/S0140525X12000660

Schilbach, L., Wohlschlaeger, A. M., Kraemer, N. C., Newen, A., Shah, N. J., Fink, G. R., & Vogeley, K. (2006). Being with virtual others: Neural correlates of social interaction. *Neuropsychologia*, *44*(5), 718–730. http://doi.org/10.1016/j.neuropsychologia.2005.07.017

Schwarzlose, R. F., Swisher, J. D., Dang, S., & Kanwisher, N. (2008). The distribution of category and location information across object-selective regions in human visual cortex. *Proceedings of the National Academy of Sciences of the United States of America*, *105*(11), 4447–4452. http://doi.org/10.1073/pnas.0800431105

Shapiro, A. G., Knight, E. J., & Lu, Z.-L. (2011). A first- and second-order motion energy analysis of peripheral motion illusions leads to further evidence of "feature blur" in peripheral vision. *PloS One*, *6*(4), e18719. http://doi.org/10.1371/journal.pone.0018719

Sinigaglia, C. (2013). What type of action understanding is subserved by mirror neurons? *Neuroscience Letters*, *540*, 59–61. http://doi.org/10.1016/j.neulet.2012.10.016

Sivak, B., & MacKenzie, C. L. (1990). Integration of visual information and motor output in reaching and grasping: the contributions of peripheral and central vision. *Neuropsychologia*, *28*(10), 1095–116. Retrieved from http://www.ncbi.nlm.nih.gov/pubmed/2267060

Spillmann, L. (2014). Receptive fields of visual neurons: The early years. *Perception*, *43*(11), 1145–1176. http://doi.org/10.1068/p7721

Spillmann, L., Ransom-Hogg, A., & Oehler, R. (1987). A comparison of perceptive and receptive fields in man and monkey. *Human Neurobiology*. http://doi.org/10.1163/_q3_SIM_00374

Strasburger, H., Rentschler, I., & Jüttner, M. (2011). Peripheral vision and pattern recognition: A review. *Journal of Vision, 11*(13), 1–82. http://doi.org/10.1167/11.5.13.Contents

Streuber, S., Knoblich, G., Sebanz, N., Bülthoff, H. H., & de La Rosa, S. (2011). The effect of social context on the use of visual information. *Experimental Brain Research, 214*(2), 273–284. http://doi.org/10.1007/s00221-011-2830-9

Sullivan, J., & Carlsson, S. (2002). Recognizing and Tracking Human Action. *Computer Vision— ECCV 2002*, 629–644. http://doi.org/10.1007/3-540-47969-4_42

Tanaka, J. W., & Curran, T. (2001). A Neural Basis for Expert Object Recognition. *Psychological Science, 12*(1), 43–47. http://doi.org/10.1111/1467-9280.00308

Tarr, M. J. (1995). Rotating objects to recognize them: A case study on the role of viewpoint dependency in the recognition of three-dimensional objects. *Psychonomic Bulletin & Review, 2*(1), 55–82. http://doi.org/10.3758/BF03214412

Tarr, M. J., & Bülthoff, H. H. (1998). Image-based object recognition in man, monkey and machine. *Cognition, 67*(1-2), 1–20. http://doi.org/10.1097/00006324-199912000-00014

Theusner, S., de Lussanet, M., & Lappe, M. (2014). Action recognition by motion detection in posture space. *The Journal of Neuroscience, 34*(3), 909–21. http://doi.org/10.1523/JNEUROSCI.2900-13.2014

Thompson, B., Hansen, B. C., Hess, R. F., & Troje, N. F. (2007). Peripheral vision: Good for biological motion, bad for signal noise segregation? *Journal of Vision, 7*(12), 1–7. http://doi.org/10.1167/7.10.12.Introduction

Thompson, J., & Parasuraman, R. (2012). Attention, biological motion, and action recognition. *NeuroImage, 59*(1), 4–13. http://doi.org/10.1016/j.neuroimage.2011.05.044

Thornton, I. M., & Vuong, Q. C. (2004). Incidental Processing of Biological Motion. *Current Biology, 14*, 1084–1089. http://doi.org/DOI 10.1016/j.cub.2004.06.025

Thornton, I. M., Wootton, Z., & Pedmanson, P. (2014). Matching biological motion at extreme distances. *Journal of Vision, 14*(3), 1–18. http://doi.org/10.1167/14.3.13.doi

Thorpe, S. J., Gegenfurtner, K. R., Fabre-Thorpe, M., & Bülthoff, H. H. (2001). Detection of animals in natural images using far peripheral vision. *The European Journal of Neuroscience, 14*(5), 869–76. http://doi.org/10.1046/j.0953-816x.2001.01717.x

Thurman, S. M., & Lu, H. (2013). Complex interactions between spatial, orientation, and motion cues for biological motion perception across visual space. *Journal of Vision, 13*(8), 1–18.

http://doi.org/10.1167/13.2.8.doi

To, M., Lovell, P. G., Troscianko, T., & Tolhurst, D. J. (2008). Summation of perceptual cues in natural visual scenes. *Proceedings. Biological Sciences / The Royal Society*, *275*(1649), 2299–308. http://doi.org/10.1098/rspb.2008.0692

To, M. P. S., Baddeley, R. J., Troscianko, T., & Tolhurst, D. J. (2011). A general rule for sensory cue summation: evidence from photographic, musical, phonetic and cross-modal stimuli. *Proceedings. Biological Sciences / The Royal Society*, *278*(1710), 1365–72. http://doi.org/10.1098/rspb.2010.1888

To, M. P. S., Lovell, P. G., Troscianko, T., & Tolhurst, D. J. (2010). Perception of suprathreshold naturalistic changes in colored natural images. *Journal of Vision*, *10*(4), 12:1–22. http://doi.org/10.1167/10.4.12

Toet, A., & Levi, D. M. (1992). The two-dimensional shape of spatial interaction zones in the parafovea. *Vision Research*, *32*(7), 1349–1357. http://doi.org/10.1016/0042-6989(92)90227-A

Traschütz, A., Zinke, W., & Wegener, D. (2012). Speed change detection in foveal and peripheral vision. *Vision Research*, *72*, 1–13. http://doi.org/10.1016/j.visres.2012.08.019

Troje, N. F. (2003). Reference frames for orientation anisotropies in face recognition and biological-motion perception. *Perception*, *32*(2), 201–210. http://doi.org/10.1068/p3392

Troje, N. F., Sadr, J., Geyer, H., & Nakayama, K. (2006). Adaptation aftereffects in the perception of gender from biological motion. *Journal of Vision*, *6*(8), 850–857. http://doi.org/10.1167/6.8.7

Troje, N. F., Troje, N. F., Westhoff, C., Westhoff, C., Lavrov, M., & Lavrov, M. (2005). Person identification from biological motion: effects of structural and kinematic cues. *Perception & Psychophysics*, *67*(4), 667–75. http://doi.org/10.3758/BF03193523

Troscianko, T. (1982). A given visual field location has a wide range of perceptive field sizes. *Vision Research*, *22*, 1361–1369.

Tynan, P. D., & Sekuler, R. (1982). Motion processing in peripheral vision: reaction time and perceived velocity. *Vision Research*, *22*(1), 61–68. Retrieved from http://www.ncbi.nlm.nih.gov/pubmed/7101752

Van Overwalle, F., & Baetens, K. (2009). Understanding others' actions and goals by mirror and mentalizing systems: A meta-analysis. *NeuroImage*, *48*(3), 564–584.

http://doi.org/10.1016/j.neuroimage.2009.06.009

Verfaillie, K. (1993). Orientation-dependent priming effects in the perception of biological motion. *Journal of Experimental Psychology. Human Perception and Performance*, *19*(5), 992–1013. http://doi.org/10.1037/0096-1523.19.5.992

Virsu, V., Rovamo, J., & Laurinen, P. (1982). Temporal Contrast Sensitivity Magnification and Cortical. *Nature*, *22*(1979).

Webster, M. a, & MacLeod, D. I. a. (2011). Visual adaptation and face perception. *Philosophical Transactions of the Royal Society of London. B*, *366*(1571), 1702–25. http://doi.org/10.1098/rstb.2010.0360

Webster, M. A. (2011). Adaptation and visual coding. *Journal of Vision*, *11*(5), 1–23. http://doi.org/10.1167/11.5.3

Webster, M. A., Kaping, D., Mizokami, Y., & Duhamel, P. (2004). Adaptation to natural facial categories. *Nature*, *428*(April), 557–561. http://doi.org/10.1038/nature02361.1.

Webster, M. A., & Leonard, D. (2008). Adaptation and perceptual norms in color vision. *Journal of the Optical Society of America A*, *25*(11), 2817–2825. http://doi.org/10.1364/JOSAA.25.002817

Whitney, D., & Levi, D. M. (2011). Visual crowding: A fundamental limit on conscious perception and object recognition. *Trends in Cognitive Sciences*, *15*(4), 160–168. http://doi.org/10.1016/j.tics.2011.02.005

Wilson, J. R., & Sherman, S. M. (1976). Receptive-field characteristics of neurons in cat striate cortex: Changes with visual field eccentricity. *Journal of Neurophysiology*, *39*(3), 512–533.

Zuiderbaan, W., Harvey, B. M., & Dumoulin, S. O. (2012). Modeling center – surround configurations in population receptive fields using fMRI. *Journal of Vision*, *12*(3), 1–15. http://doi.org/10.1167/12.3.10.Introduction

## Declaration of the Contribution of the Candidate

This Thesis is presented in the form of a collection of manuscripts that are, at the time of thesis submission, either published or prepared for publication. Details about these manuscripts are presented in the following.

1. Fademrecht, L., Bülthoff, I. & de la Rosa, S. (2016).

   Action recognition in the visual periphery. *Journal of Vision*, 16(3), 1-14:

   Design, stimulus generation, experimental work and analysis of the study have predominantly been developed and finalized by the candidate. The co-author's role was that of supervision in giving advice, offering knowledge and criticism, and revising the manuscript.

2. Fademrecht, L., Bülthoff, I. & de la Rosa, S. (2016).

   Viewpoint dependent action recognition processes in the visual periphery (prepared for submission):

   Design, stimulus generation, experimental work and analysis of the study have predominantly been developed and finalized by the candidate. The co-author's role was that of supervision in giving advice, offering knowledge and criticism, and revising the manuscript.

3. Fademrecht, L., Bülthoff, I., Barraclough, N. E. & de la Rosa, S. (2016).

   Measuring perceptive field sizes of action sensitive perceptual channels (prepared for submission):

   Design, stimulus generation, experimental work and analysis of the study have predominantly been developed and finalized by the candidate. The co-author's role was that of supervision in giving advice, offering knowledge and criticism, and revising the manuscript.

4. Fademrecht, L., Nieuwenhuis, J., Bülthoff, I., Barraclough, N. E. & de la Rosa, S. (2016).

Action adaptation in a crowded environment (submitted, *Frontiers in Psychology*):

Design, stimulus generation, experimental work and analysis of the study have predominantly been developed and finalized by the candidate. The co-author's role was that of supervision in giving advice, offering knowledge and criticism, and revising the manuscript.

Parts of this work was also presented at the following conferences:

1. Fademrecht, L., de la Rosa, S. (2016). Towards action recognition in the loop. Talk to be presented at the 28th Annual Convention of the Association for Psychological Science, Chicago, USA.

2. Fademrecht, L., Nieuwenhuis, J., Bülthoff, I., Barraclough, N.E., de la Rosa, S. (2016). Does a crowded environment influence action perception? Poster presented at the 16th Annual Meeting of the Vision Sciences Society, St. Pete Beach, USA.

3. Fademrecht, L., Nieuwenhuis, J., Bülthoff, I., Barraclough, N.E., de la Rosa, S. (2015). The spatial extent of action sensitive perceptual channels decrease with visual eccentricity. Poster presented at the 45th Annual Meeting of the Society for Neuroscience, Chicago, USA.

4. Fademrecht, L., Bülthoff, I., Barraclough, N.E., de la Rosa, S. (2015). Seeing Actions in the fovea influences subsequent recognition in the visual periphery. Poster presented at the European Conference on Visual Perception, Liverpool, UK.

5. Fademrecht, L., Bülthoff, I., de la Rosa, S. (2015). Recognition of static and dynamic actions in the visual periphery. Poster presented at the 15th Annual Meeting of the Vision Sciences Society, St. Pete Beach, USA.

6. Fademrecht, L., Bülthoff, I., de la Rosa, S. (2014). A Matter of perspective: Action recognition depends on stimulus orientation in the periphery. Poster presented at the European Conference on Visual Perception, Belgrade, Serbia.

7. Fademrecht, L., Bülthoff, I., de la Rosa, S. (2014). Peripheral vision and action recognition. Poster presented at the 6th International Conference on Brain and Cognitive Engineering, Tübingen, Germany

8. Fademrecht, L., Bülthoff, I., de la Rosa, S. (2014). Influence of eccentricity on action recognition. Poster presented at the 14th Annual Meeting of the Vision Sciences Society, St. Pete Beach, USA.

# 2 STUDY I: ACTION RECOGNITION IN THE VISUAL PERIPHERY

Laura Fademrecht, Isabelle Bülthoff, Stephan de la Rosa

Max Planck Institute for Biological Cybernetics, Tübingen, Germany

## 2.1 Abstract

Recognizing whether the gestures of somebody mean a greeting or a threat is crucial for social interactions. In real life, action recognition occurs over the entire visual field. In contrast, much of the previous research on action recognition has primarily focused on central vision. Here our goal is to examine what can be perceived about an action outside of foveal vision. Specifically, we probed the valence, as well as first level and second level recognition of social actions (handshake, hugging, waving, punching, slapping, and kicking) at 0° (fovea/fixation), 15°, 30°, 45° and 60° of eccentricity with dynamic (Experiment 1) and dynamic and static (Experiment 2) actions. To assess peripheral vision under conditions of good ecological validity, these actions were carried out by a life-size human stick-figure on a large screen. In both experiments, recognition performance was surprisingly high (over 66% correct) up to 30° of eccentricity for all recognition tasks and followed a nonlinear decline with increasing eccentricities.

## 2.2 Introduction

Recognition of human actions is crucial for social interaction. So far, most studies have investigated the visual mechanisms underlying action recognition at fixation (central vision) and largely ignored peripheral vision. However, in real life we are aware of actions happening not only in central vision but also in the visual periphery. For example, in a conversational setting we are still aware of our partner's hand movements despite focusing on his face. The purpose of the present study is to examine the recognition of social actions throughout the visual field, that is, in the central and peripheral regions of the retina.

Many of the studies investigating visual recognition of bodily movements within the central vision field have shown that humans are able to read a large range of information from biological motion (see Blake & Shiffrar, 2007 or Giese, 2013 for comprehensive reviews), for example, the actor's identity (Cutting & Kozlowski, 1977; Loula et al., 2005), intention (Runeson & Frykholm, 1983), or sex (Barclay et al., 1978; Kozlowski & Cutting, 1977). Yet, everyday social interactions also require humans to be exquisite at recognizing actions. For example, the generation of an appropriate complementary action requires the observer to determine whether the interaction partner is carrying out a punch or a handshake. Only a few studies have investigated the recognition of social actions. They have shown that their recognition is sensitive to the temporal synchrony and the semantic relationship of the interaction partners actions (Manera et al., 2011; Neri et al., 2006). Additionally, social action recognition is also sensitive to viewpoint (de la Rosa et al., 2013) and to the social context in which an action is embedded (de la Rosa, Streuber, et al., 2014; Streuber, Knoblich, Sebanz, Bülthoff, & de La Rosa, 2011). Moreover, we can recognize the same action on several cognitive abstraction levels (first level: e.g. handshake; second level: e.g. greeting) (de la Rosa, Choudhery, et al., 2014).

Action recognition in the visual periphery has received little attention. The few existing studies, which all used point light stimuli, have mainly focused on the detection and the direction discrimination of locomotive actions (e.g. walking, running) at eccentricities up to 12° (near periphery). Their results show that these actions can be readily detected at these eccentricities, although there was always a disadvantage in the periphery compared to central vision (Ikeda et al., 2005, 2013; B. Thompson et al., 2007).

There are several reasons to assume that the role of peripheral vision with regards to action recognition goes beyond the detection of biological motion and the discrimination of the direction of an action. Previous research suggests that at least two other important aspects of an action could be detected in the periphery, namely we can judge its emotional valence and classify it at various abstraction levels. As for valence, face recognition research suggests that affective face information can be readily recognized in the visual periphery (Bayle et al., 2009; Bayle, Schoendorff, Hénaff, & Krolak-Salmon, 2011; Rigoulot et al., 2012). With regards to actions, the recognition of their emotional valence in the visual periphery would, for example, allow an early detection of a threatening action. In terms of cognitive abstraction levels, previous research has shown that action categorization occurs on several abstraction levels (de la Rosa, Choudhery, et al., 2014). For instance, participants could describe a handshake action as a greeting or on a more detailed level as a handshake. The former is referred to as recognition at the second level and the latter as recognition at the first level. These different recognition levels result in different levels of recognition performance. For the description of these recognition levels we refer to the specificity of the actions. That means that for the more specific group of actions (e.g. handshake) we use the term 'first level' and for the more general group of actions (e.g. greeting) we will use the term 'second level'. In congruence with the object recognition literature, where the second level is described as basic level and the first level as subordinate level (Rosch et al., 1976), actions are recognized more accurately and faster at the second than at the first level (de la Rosa, Choudhery, et al., 2014).

In two experiments we examined the visual recognition of social actions in the periphery. In Experiment 1 we examined the recognition of dynamic actions with respect to their valence, first and second level of recognition. In Experiment 2, we investigated valence, first and second level recognition for static and dynamic actions.

## 2.3 Experiment 1

Our aim was to examine valence (positive vs. negative), first level (e.g. handshake) and second level (e.g. greeting) action recognition over a large portion of the visual field (fixation, near periphery and far periphery). To mimic realistic viewing scenarios, we used dynamic actions (i.e. movies) and kept the size of the life-size actor constant across the visual field instead of adjusting the stimulus size to compensate for the reduced resolution in the periphery (cortical magnification: Cowey and Rolls 1974; Daniel and Whitteridge 1961; Rovamo and Virsu 1979).

## 2.3.1 Methods

*Participants:* We recruited 45 participants (18 males, 27 females) from the local community of Tübingen. All participants received monetary compensation for their participation. Their age ranged from 19 to 53 years (mean: 27.1). The participants had normal vision or corrected their visual acuity using contact lenses. Participants gave their written, informed consent form prior to the experiment. The study was conducted in line with Max Planck Society policy and had been approved by the University of Tübingen ethics committee.

*Apparatus*: Stimuli were presented on a large panoramic screen with a half-cylindrical virtual reality projection system (Figure 1). The wide screen display amounts to 7 m in diameter and 3.2 m in height (230° horizontally, 125° vertically). Six LED DLP projectors (1920x1200, 60Hz) (EYEVIS, Germany) were used to display the stimuli against a grey background on the screen. The geometry of the screen can be described as a quarter-sphere. The visual distortions of the display caused by the curved projection screen were compensated with the use of warping technology software (NVIDIA, Germany). At all eccentricities the screen was 2 m away from the subject and the stimuli were presented at a virtual distance of 3 m. Participants sat on a stool in front of a desk in the middle of the quarter-spherical arena. They placed their head on a chin and forehead rest which was mounted on the desk (see Figure 1). During each experimental trial they were required to keep their eyes focused on a grey fixation cross placed on the screen straight ahead of them. This position of the cross was defined as 0° position. An eye tracker (Eyelink II, SR Research Ltd., Canada) was used to control for their eye movements. When the stick figure was presented at 0°, it was presented behind the cross. The Unity 3D (Unity Technologies, USA) game engine in combination with a custom written control script was used to control the presentation of the stimuli and to collect responses.

*Figure 1: Projection system: Large panoramic screen with chair, chinrest and eye tracker.*

*Stimuli:* Actions were recorded via motion capture using Moven Suits (XSens, Netherlands). The Xsens MVN Suits consists of 17 inertial and magnetic sensor modules, which are placed in an elastic lycra suit worn by the actor. The sampling rate was 120 Hz. Three actions with positive emotional valence (handshake, hugging and waving) and three actions with negative valence (slapping, punching and kicking) were acted out each by six different lay actors (three male, three female). Every action was repeated six times by each actor, leading to 216 stimuli in total. The actions lasted between 800 and 1500 ms and each action started with the actor standing in a neutral position, i.e. with their arms aligned with the body, and ended with the peak frame of the action. The peak frame of an action was determined as the point in time just before the actor started moving back to the neutral position.

The motion data was mapped onto a grey life-size 'stick figure avatar' (avatar height: 170 cm, around 32° visual angle). The position of the stick figure avatar in the visual field (on the screen) was determined by the position midway between both hips. The stimuli were always oriented toward the participant at any position on the screen and were always presented along the same latitude (i.e. on the same horizontal axis). A stick figure (see Figure 1) was used instead of a full-fleshed avatar in order to prevent any other visual cues like appearance or gender from influencing participant's decisions. Furthermore, using a stick figure had the advantage that we did not have to record facial movement information (e.g. expression and gaze) and hand and feet motions. We favored the use of a stick figure over a dynamic point-light display because the sparse structure of the latter might unduly hinder recognition because of the decreasing spatial resolution of the visual system toward the periphery.

*Procedure and Design:* At the beginning of the experiment participants were informed about the following experimental procedure (Figure 2). Each trial began with the presentation of a fixation cross and the eye tracker started to record the eye movements. Participants were told to fixate the cross (trials with a gaze shift larger than 2° were discarded from all analyses). The stick figure appeared at one of nine positions (-60°, -45°, -30°, -15°, 0°, 15°, 30°, 45° or 60°) in the participant's visual field. Participants were instructed to answer one of the following three questions in a between-subject design. They answered either the question "What action did you see?", in order to identify the action they had seen (first level task), or they answered the question "Was the action a greeting or an attack?", meaning that they categorized the actions at the second level (second level task), or they answered the question "Was the action positive or negative?" to evaluate the emotional valence of the viewed action (valence task). There were two answer options in the valence task (i.e. positive or negative) and in the second level task (i.e. greeting or attack). There were six answer options in the first level task (handshake, hugging, waving, kicking, punching or slapping). Participants were asked to answer as quickly and as accurately as possible. The answer could be given as soon as the stick figure appeared on the screen. When participants did not respond before the end of the animation sequence, a prompt appeared on the screen, displaying the question and the pre-defined response keys on a keyboard (1 or 0 on the keyboard for the second level and the valence task; 1, 2, 3, 8, 9, 0 on the keyboard for the first level task). Three of the actions had a positive emotional valence (handshake, hugging, waving) and three had a negative emotional valence (kicking, punching, slapping). Each of these six actions was presented 100 times. We manipulated the eccentricity of the stick figure, so that it appeared at 0°, 15°, 30°, 45° or 60° away from the fixation cross (Figure 1). The stick figure appeared randomly either on the left or on the right side of the screen for positions other than 0°. The actions (and hence the valence) and their positions on the screen were counterbalanced within each task with each action presented 20 times at each location. This resulted in a total of 600 trials per task (20 repetitions x 5 position x 6 actions). In 600 trials the 216 stimuli were shown 2.8 times on average. Actions and positions were in a different random order for every participant.

Each recognition task was performed by a separate group of 15 participants. Hence recognition task was a between-subjects factor and position, action and valence were within-subjects factors. At the beginning of an experiment participants received a short training in order to get used to the setup and the task. In the second level and valence tasks this training lasted for 10 trials, in the first level task participants received a longer training phase of 20 trials to learn the response key – action associations. The stimuli used in the training trials were different from the stimuli in the test trials.

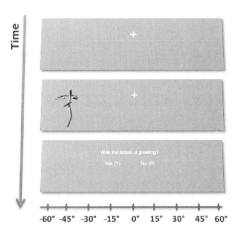

*Figure 2: Sequence of events within a trial.*

## 2.3.2 Results

Accuracy and reaction times served as measures for recognition performance and their results are presented here separately. In less than 0.8% of the trials participants failed to fixate the cross in the middle of the screen. These trials were discarded. For the analysis we collapsed the data of the right and left visual field. Reaction time data was filtered for outliers and reaction times below 200 ms and above 4000 ms were discarded (0.6% of the trials).

## Reaction times

Only reaction times for correct responses were considered in this analysis. The mean reaction time overall was 1181 ms ($SE$ = 16 ms). Participants' reaction times increased with eccentricity for each recognition task and were task dependent (Figure 3). We ran a mixed-effects model with recognition task (first level, second level, valence) and eccentricity (0°, 15°, 30°, 45°, 60°) as fixed factors and participant as random factor. The slope for eccentricity was fitted in a by-participant fashion. The results showed a significant main effect of eccentricity ($F(1, 177)$ = 184.51, $p$ < .0001) and a significant main effect of recognition task ($F(2, 42)$ = 16.56, $p$ < .001). The results suggest that reaction times were dependent on the stimulus position in the visual field and on the recognition task. We examined the effect of task with pairwise t-tests using a Bonferroni correction for multiple comparisons. These showed that participants answered faster in the valence recognition task than in the second level task (valence vs. second level: $t_{paired}$ = 5.42, $df$ = 147.57, $p$ < .001). They showed the longest reaction times in the first level task (second level vs. first level: $t_{paired}$ = 7.12, $df$ = 130.97, $p$ < .001). This worse performance in the first level task might be due to the larger number of response options (6 response options) compared to the valence and the second level task (2 response options). The two-way interaction was significant ($F(2, 177)$ = 3.51, $p$ = .032), which shows that for each recognition task reaction times increased differently with eccentricity. The significant interaction between recognition task and eccentricity was examined using Dunnett's test. We were interested in the position at which recognition performance in the periphery started to differ significantly from foveal vision. We therefore compared all peripheral positions to 0° eccentricity. For the second level and the valence task reaction times at fixation and in the periphery did not differ significantly from each other up to and including 45° eccentricity (all $p$ values higher than .1). Thus the only significant difference was between 0° and 60° (second level: $t_{paired}$ = 4.07, $p$ < .001; valence: $t_{paired}$ = 3.29, $p$ < .01), indicating that there was no significant increase of reaction times before testing at 60° eccentricity. For the first level task there was even no significant difference to the reaction times at fixation for all tested eccentricities (all *p* values higher than .1).

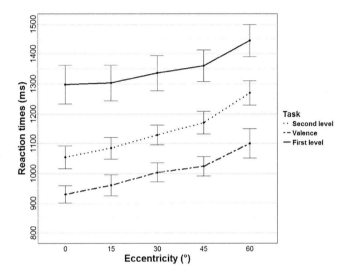

*Figure 3: Means and standard errors of reaction times for the three recognition tasks as a function of eccentricity.*

**Accuracy**

To account for the fact that the first level task had six response options while the second level and the valence tasks had only two, we corrected the accuracy results statistically for guessing according to Macmillan and Creelman (2005, page 252):

$$c = \frac{[m \cdot p(c) - 1]}{(m - 1) \cdot 100}$$

This formula gives the accuracy corrected for guessing in percent c. The parameter p(c) is the probability of a correct response, m is the number of response options in a given task (for the valence and the second level task m = 2, for the first level task m = 6).

The mean accuracy was well above chance level for each task over all tested eccentricities (overall accuracy: $M = 0.88$, $SE = 0.005$). Figure 3 shows that recognition performance decreased with eccentricity in all tasks and that accuracy was lower for the first level task than for the second level and the valence task. Furthermore, the decline of performance with eccentricity seems stronger for the first level task than for the two other tasks. A mixed-effects model with recognition task (first level, second level, valence) and eccentricity (0°, 15°, 30°, 45°, 60°) as fixed factors and a random slope for eccentricity that was fitted in a by-participant fashion revealed a significant main effect of eccentricity ($F(1, 177) = 57.32$, $p < .001$), indicating that recognition performance decreased with increasing eccentricity of the stimulus position in the visual field. There was also a significant main effect of recognition task ($F(2, 42) = 29.47$, $p < .0001$), showing that participants' accuracy depended on the task requirements. The significant two-way interaction of task and eccentricity ($F(2, 177) = 9.26$, $p < .0001$) suggests that eccentricity affected recognition in the three tasks differently. In Figure 4 we plotted the 95% confidence intervals. This illustrates the significant differences between the three tasks at the different positions. For 0° and 15° positions there is no performance difference between the tasks, whereas from 30° onwards the first level task always leads to lower accuracy rates than the second level and the valence tasks whose data did not differ from each other at any position. Between the valence and the second level task we found no difference in accuracy. We used a Dunnett's test for each recognition task to compare the recognition performance at the peripheral positions with the performance at fixation. In all three recognition tasks the recognition performance up to 45° eccentricity did not differ significantly from the performance at fixation, thus indicating that the decline of recognition performance starts after 45° (Dunnett's test was only significant for comparisons between 0° and 60° in all recognition tasks: valence $t = -2.57$, $p = 0.04$; second level $t = -3.5$, $p < 0.01$; first level $t = -5.12$, $p < 0.001$). Figure 4 in combination with the statistical analysis indicates a nonlinear relationship of recognition performance with eccentricity for all three recognition tasks.

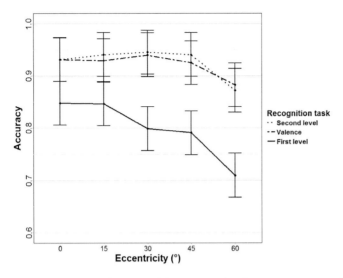

*Figure 4: Means and 95% confidence intervals of response accuracy for the three recognition tasks, as a function of eccentricity.*

## 2.3.3 Discussion

In this experiment we tested human action recognition from central vision up to 60° eccentricity in three different recognition tasks (first level, the second level and the emotional valence). Reaction times in all three tasks increased with eccentricity. Moreover, participants were fastest in the valence task and slowest in the first level task, thus confirming previous findings (de la Rosa, Choudhery, et al., 2014). The accuracy data also indicate that all tasks get harder with increasing eccentricity. The better recognition performance observed in foveal vision for second level than for first level categorization therefore seems to extend to peripheral vision. Additionally, participants show the best performance in the valence task. For this task accuracy is as high as in the second level task while reaction times are shorter than in both other tasks. The significant interaction between task and eccentricity seems to be owed to the steeper decline in recognition performance in the first level task compared to the second level and valence tasks. A simple explanation for this pattern is that more detailed visual information is needed for the recognition at the first level. This type of information might not be accessible due to the sparse resolution in the visual periphery.

Remarkably, accuracy declined nonlinearly with increasing eccentricity but remained above chance level for all tested eccentricities. The former was partly unexpected since previous research examining recognition of static objects in the periphery (Jebara et al., 2009; Thorpe et al., 2001) reported a linear decline of recognition performance with eccentricity. Can the motion energy in our dynamic stimuli account for this difference? The literature about motion perception in peripheral vision describes a rather linear decline with eccentricity for first- and second-order motion. Few studies describe a nonlinear relationship (Tynan & Sekuler, 1982). In this study the authors measured motion detection thresholds up to 30° eccentricity and found a nonlinear increase of detection thresholds. However, it is important to note that the motion patterns induced by limb movement in our stimuli are more complex (e.g. they consist of many more movement orientations in 3D space) than the ones employed in previous studies with low-level motion stimuli (first- and second-order motion). Hence, we cannot rule out that participants might have relied on the additional motion cues in our stimuli to maintain a high recognition performance far into the periphery. If motion cues were responsible for the nonlinear decline in our study then presenting static action images instead of action movies should result in a linear decline of performance with eccentricity. To test this hypothesis we conducted a second experiment where we compared the recognition of action movies with the recognition of static representations of actions (images).

## 2.4 Experiment 2

Experiment 2 investigated the influence of motion information on the recognition of social actions in the visual periphery. We changed the experimental methods to overcome two important shortcomings of Experiment 1. First, the three recognition tasks in Experiment 1 had different numbers of response options (six response options in the first level task and two response options in the valence and second level tasks). The larger number of response options in the first level task might have been responsible for the slower reaction times and lower accuracy in that task. To avoid this problem, we changed the experimental design in such a way that all recognition tasks had two response options. Specifically, participants were presented with one action at a time and had to indicate whether the presented action matched a predefined first level (e. g. punching), second level (e.g. greeting) or valence (e.g. positive). Hence, all three tasks relied on a yes-no task. Yes-no tasks have been frequently used for the investigation of visual object categorization and it has been shown that switching from a n-alternative forced choice (n-AFC) task (with n>2) to a yes-no task does not change the overall pattern of the results in object categorization tasks (de la Rosa, Choudhery, & Chatziastros, 2011; Grill-Spector & Kanwisher, 2005). Hence, in Experiment 2 we switched to a yes/no task for obtaining a more fair comparison of the different recognition tasks. Second, positive and negative valence actions were associated with different motion energies in the stimuli of Experiment 1. Therefore, participants might have defaulted to a recognition strategy that relied on simply assessing the amount of motion energy in the second level and valence recognition task. Therefore we added distractor actions in Experiment 2 that had motion energy similar to the actions described in Experiment 1 but did not show any meaningful actions. To create these actions we remapped the arm motion onto the legs and vice versa.

## 2.4.1 Methods

We used the same methods as in Experiment 1 except for the following.

*Participants:* We recruited 19 participants from the local community of Tübingen (9 males, 10 females). The age ranged from 20 to 39 years (mean: 26.1).

*Stimuli:* In addition to the action stimuli of Experiment 1, we created distractor stimuli in the following two ways: Either we remapped the left and right arm movements onto the left and right legs and vice versa or we mapped the left leg movement onto the right arm and vice versa. We will refer to these distractor stimuli as remapped distractor stimuli. Importantly, no action could be recognized from these actions thereby rendering them meaningless.

*Procedure and Design:* We measured the recognition of dynamic and static actions in two separate experimental sessions (testing order was counterbalanced across participants). Each experimental session consisted of ten experimental conditions (two for the valence task, two for the second level task and six for the first level task). At the beginning of each experimental condition participants received verbal instructions about the question they had to answer in that condition. Each question probed the recognition of a different action target. The questions "Was the action positive?" or "Was the action negative?" probed valence recognition and the questions "Was the action a greeting?" and "Was the action an attack?" measured second level recognition. We used the following six questions to measure first level recognition: "Was the action a handshake?", "Was the action a hug?", "Was the action a wave", "Was the action a kick?", "Was the action a punch?" and "Was the action a slap?". Participants always had the response options "yes" (for the target action) and "no" (for non-targets) for each question in the static and the dynamic experimental sessions. For remapped distractors, a correct response was "no" to all questions. An experimental trial started with the presentation of the fixation cross at 0° and the stick figure avatar appeared at one of the nine positions in the participant's visual field as described in the first experiment. The answer could be given as soon as the stick figure appeared on the screen. When participants did not respond before the end of the animation sequence, a prompt appeared on the screen, displaying the question and the pre-defined response keys on a keyboard. In each experimental condition, 50% of the trials showed the target action and the remaining trials showed distractors. 50% of the distractors were remapped distractor stimuli and the remaining were non-target actions that were not remapped. For example if the target was "positive actions", 50% of the experimental trials showed the three positive actions as target (hugging, waving, handshake), 25% showed remapped distractors derived from the positive actions, and 25% of the trials showed the three actions with negative valence (kicking, punching, slapping). The testing order of the ten experimental conditions was randomized across participants. In each condition, each target action was presented 12 times at each location of each hemi-field. The valence and second level tasks had 480 trials ((12 repetitions x 2 target presence (present vs. absent) x 5 locations x 2 hemi- fields (left or right side of the visual field) x 2 questions (for example: "Was the action positive?"/"Was the action negative?")). The first level task had 80 trials for each of the six questions (for example: "Was the action a handshake?" and so on) (4 repetitions x 2 target presence (present vs. absent) x 5 locations x 2 hemi-fields). The two experimental sessions probing the recognition of static and dynamic actions were carried out on different days. This resulted in a total of 2880 trials per participant ((2 x 480 trials in the second level and valence condition + 480 trials in the first level condition) x 2 sessions). Recognition task, position and motion type (static vs dynamic) were within-subjects factors.

## 2.4.2 Results

We calculated the sensitivity (d') according to Macmillan & Creelman (2005) as a measure of recognition performance. 1% of the trials were excluded due to deviation from fixation. Reaction times for correct target identification and sensitivity are evaluated separately. Reaction time data was filtered for outliers and reaction times below 200 ms and above 3500 ms were discarded (0.2% of the trials).

**Reaction times**

Participants' reaction times increased with eccentricity for each task and both motion types (Figure 6). The mean reaction time for the whole experiment was 894 ms ($SE = 2$ ms) calculated over all data in all tasks. We analyzed only the reaction times for correct target identification. We used a mixed-effects model with recognition task (first level, second level, valence) and eccentricity (0°, 15°, 30°, 45°, 60°) and motion type (static, dynamic) as fixed factors and a random slope for eccentricity that was fitted in a by-participant fashion to investigate the reaction times. In order to examine the relationship between the reaction time performance and eccentricity we treated eccentricity as a continuous variable. We found a significant main effect of recognition task ($F(2, 538) = 45.65$, $p < .001$). In the first level task, participants answered with shorter reaction times ($M_{RT} = 843$ ms, $SE = 1$) than in the second level task ($M_{RT} = 883$ ms, $SE = 1$) and in the valence task ($M_{RT} = 937$ ms, $SE = 2$). The significant main effect of eccentricity ($F(1, 538) = 87.51$, $p < .001$) indicates that participants reaction times were dependent on the stimulus position and increased with increasing eccentricity. The motion type had a significant main effect on the reaction times as well ($F(1, 538) = 6.35$, $p = 0.01$), participants showed shorter reaction times for the static condition than for the dynamic condition, although this difference did depend on the recognition task, as the significant interaction between motion type and recognition task ($F(2, 538) = 3.86$,

$p$ = 0.02) shows. Pairwise t-tests, using a Bonferroni correction for multiple comparisons, showed that reaction times for dynamic and static stimuli differed significantly from each other only in the first level task ($t_{paired}$ = 4.01, $df$ = 94, $p$ < .001). All other comparisons and interactions were non-significant (all $p$ values > .05).

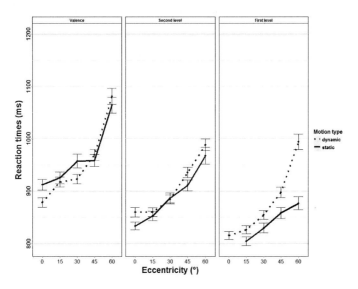

*Figure 6: Means and standard errors of the reaction times as a function of eccentricity for static and dynamic action stimuli in the three recognition tasks (Note that the scales of the y-axis differ in Experiment 1 and 2)*

**Sensitivity (d')**

A cursory look at the graph (Figure 7) indicates that, for dynamic stimuli, participants were always clearly able to discriminate between target and distractor actions at all probed locations as indicated by d' values higher than 0 while for static stimuli this was true only up to 30° eccentricity. A mixed-effects model with recognition task (first level, second level, valence) and eccentricity (0°, 15°, 30°, 45°, 60°) and motion type (static, dynamic) as fixed factors and a random slope for eccentricity that was fitted in a by-participant fashion shows a significant main effect of recognition task ($F(2, 540)$ = 350.46, $p$ < .0001). In the first level task ($M_{d'}$ = 1.66; $SE$ = 0.06) participants reached significantly higher d' values than in the second level ($M_{d'}$ = 0.67; $SE$ = 0.05) and in the valence task ($M_{d'}$ = 0.66; $SE$ = 0.05; t-test: valence vs. first level task $t_{paired}$ = 12.83, $df$ = 374.84, $p$ < .001; second level vs. first level task $t_{paired}$ = 13.16, $df$ = 365.22, $p$ < .001; second level vs. valence task $t_{paired}$ = -0.24, $df$ = 374.48, $p$ < .811). This finding indicates a better recognition performance in the first level task than in the two other recognition tasks. The main effect of eccentricity was significant as well ($F(1, 540)$ = 100.55, $p$ < .001). The mean d' averaged over all three tasks and the two conditions is decreasing with eccentricity, starting with a mean d' of 1.38 ($SE$ = 0.07) at fixation and ending with a mean d' of 0.3 ($SE$ = 0.07) at 60°. We examined the main effect of eccentricity using Dunnett's test. Sensitivity values at all peripheral positions were compared to that at fixation. We found a significant difference between 0° and 45° ($t_{paired}$ = -5.55, $p$ < .001) and as well for 0° and 60° ($t_{paired}$ = -10.64, $p$ < .001). These results indicate that the decline of recognition performance starts after 30° eccentricity. The significant main effect of motion type ($F(1, 540)$ = 456.86, $p$ < .001) shows that response accuracy is also sensitive to the experimental condition (static or dynamic), resulting in a mean d' of 0.62 ($SE$ = 0.05) for the static condition and a mean d' of 1.37 ($SE$ = 0.05) in the dynamic condition. Thus, dynamic target stimuli are better discriminated from distractors than static target stimuli. All higher order interactions were non-significant (all $p$ values > .05), including the two-way interaction between motion type and eccentricity ($F(1, 540)$ = 0.58, $p$ = .45).

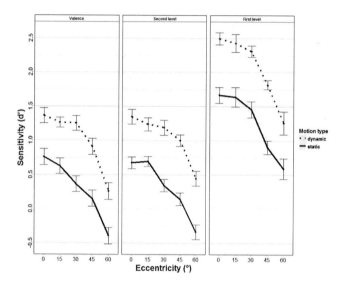

*Figure 7: Means and standard errors of sensitivity (d') as a function for eccentricity for static and dynamic action stimuli in the three recognition tasks.*

**Assessing the change of recognition performance with eccentricity**

In order to assess whether the performance changed linearly with eccentricity we examined the relationship between recognition performance and eccentricity more formally. Specifically, we fitted a power law function to the performance data of Experiment 2 for each participant, separately for the dependent variable (RT and d') and motion type (dynamic and static). The reasons for using a power law function were twofold. First, power laws have been shown to well describe relationships between physical properties and their perception (e.g. Steven's power law). Second, these functions give also the opportunity of a linear fit (exponent would then be 1), therefore we could directly test whether the performance declines in a linear or a nonlinear fashion with eccentricity. The fits were carried out by means of the 'gfit' function in MATLAB. We fitted the following power law function:

$$f(x) = w \cdot a \cdot x^b + c$$

The parameter w indicates whether the change in performance was increasing (w = 1) or decreasing (w = -1) with eccentricity. We set w = -1 for fitting the d' data to describe the decrease of d' with eccentricity. Likewise, we set w = 1 for the fitting of the reaction time data to describe the increase in RT with larger eccentricities. The parameter 'a' is the slope of the function and scales the function along the y axis. 'b' is the exponent and defines the type of relationship: for a linear relationship we expect b = 1 and for nonlinear relationships b ≠ 1. 'c' is the intercept of the curve and is a measure of recognition performance at fixation. Parameters a, b and c were free to vary.

We present the results from this analysis for reaction times and d' separately.

**Reaction times**

On average, the power law functions fit the data well both in the static and dynamic condition (mean $R^2$ 0.76 and 0.92, respectively). To assess the linearity of the performance decrease with eccentricity, we tested the exponents against 1. The mean exponents in the dynamic ($M_{exp}$ = 2.75, SE = 0.35) and the static condition ($M_{exp}$ = 2.90, SE = 0.52) were significantly different from 1 (dynamic: $t_{paired}$ = 4.94, $df$ = 18, $p$ < .001; static: $t_{paired}$ = 3.62, $df$ = 18, $p$ = .002), suggesting a nonlinear relationship between reaction time and eccentricity. There was no significant difference between the exponents of dynamic and static conditions ($t_{paired}$ = 0.35, $df$ = 18, $p$ = .73). The mean values for the intercept 'c' and the slope 'a' are listed in Table 1, as well as the $R^2$ values for the different conditions.

*Table 1: Mean parameters for the power law function fitted to each participant's individual reaction time data.*

| Motion type | Exponent 'b' | Slope 'a' | Intercept 'c' | $R^2$ |
|---|---|---|---|---|
| Static | 2.90 | 89.68 | 841.94 | 0.76 |
| Dynamic | 2.75 | 119.47 | 861.95 | 0.92 |

**Sensitivity (d')**

The power law function fitted the d' data well in both conditions ($R^2$ in the static condition: 0.83; $R^2$ in the dynamic condition: 0.81). The mean exponents in both the dynamic ($M_{exp}$ = 3.09, $SE$ = 0.17) and static condition ($M_{exp}$ = 3.02, $SE$ = 0.14) were significantly different from 1 (dynamic: $t_{paired}$ = 12.36, $df$ = 18, $p < .001$; static: $t_{paired}$ = 13.91, $df$ = 18, $p < .001$). Therefore, this result suggests that d' changes in a nonlinear fashion with eccentricity. There was no significant difference ($t_{paired}$ = 0.41, $df$ = 18, $p$ = .68) between the mean exponents of the static and dynamic conditions. The mean values for the intercept 'c' and the slope 'a', as well as the $R^2$ values are given for the different conditions in Table 2.

*Table 2: Mean parameters for the power law fitted on each participant's individual d' data.*

| Motion type | Exponent 'b' | Slope 'a' | Intercept 'c' | $R^2$ |
|---|---|---|---|---|
| Static | 3.02 | 1.40 e-05 | 0.95 | 0.83 |
| Dynamic | 3.09 | 1.29 e-05 | 1.70 | 0.81 |

## 2.4.3 Discussion

By using dynamic and static action stimuli in Experiment 2, we examined whether it was the presence of motion in the stimuli in Experiment 1 that had led to the nonlinear decline of recognition performance with increasing eccentricity. While, participants did not differ in terms of reaction times between static and dynamic action stimuli in the valence and the second level task, participants showed a higher sensitivity for dynamic than for static actions over all positions in the visual field and all three tasks. Importantly, the absence of an interaction between motion type and eccentricity in our analysis indicates that recognition of static and dynamic actions declines in a similar fashion with eccentricity. A significant difference between the response times to static and dynamic stimuli is observed only in the second level recognition task, which might be explained by a flooring effect in the dynamic action condition. Dynamic actions were presented as videos (1 to 2 s long) while our static stimuli presented each action as one image extracted at the peak of the action (see the method section for more details). Hence, important action information was immediately visible for static actions but not for dynamic actions and allowed faster responses. The fastest response time recorded (800 ms) for videos reflects the minimum time needed to recognize an action in our dynamic stimuli while response time can be much shorter with static stimuli. In line with this idea, a minimum reaction time of 800 to 900 ms has also been found in a study of de la Rosa and colleagues (de la Rosa et al., 2013) using dynamic movies of other actions. Hence, we think that the larger difference in response time between static and dynamic actions found in the first level task is simply owed to a flooring effect for dynamic but not static actions. Together these findings indicate that discriminating targets from their distractors was easier for dynamic than static stimuli at all positions in the visual field. In the dynamic condition participants were clearly able to discriminate between target actions and distractors, even in the most peripheral positions. It is important to note that, they could discriminate target actions from distorted meaningless actions that were based on the motion data of the target actions. This shows that participants did not use the stimulus' motion energy as a cue to discriminate, for example, positive from negative actions. We assessed more formally the relationship between recognition performance (RT and d') and eccentricity for dynamic and static stimuli using a power law function. This analysis indicates that both RT and d' change in a nonlinear fashion with eccentricity. Moreover, this nonlinear decline did not vary with motion type, a nonlinear decline was observed when using dynamic as well as static stimuli. Hence, the nonlinear decline of action recognition performance with eccentricity was unlikely due to the motion information present in the dynamic stimuli since a nonlinear decline was also observed when using static stimuli.

What other factors might explain the nonlinear decline of action recognition performance in the visual periphery? Unfortunately, most of the behavioral studies examining recognition performance in the periphery reported a linear decline and therefore provide little insights into the nonlinear nature of our results. One hypothesis that we are currently assessing in our lab is whether the perceptual field size of action sensitive channels can account for the nonlinear decline of action recognition performance in the visual periphery. Specifically, one way to account for these results is that foveal perceptual channels extend into the periphery thereby increasing recognition performance there. However, the reason for observing a nonlinear decline instead of a linear one is still unsettled and what it means in terms of underlying perceptual mechanisms remains to be elucidated.

## 2.5 General discussion

Our results demonstrate that participants are remarkably good at recognizing actions in the periphery. Recognition performance for social actions remained above chance level up to 60° for dynamic action stimuli. Recognition of static action stimuli remained reliable up to 30° eccentricity, which indicates that up to that level of eccentricity participants do not only rely on motion information when recognizing actions. Moreover, participants were not only able to tell the valence level of an action shown dynamically up to 60°, but also its first level and second level. Hence, participants recognize much more than the emotional gist of an action even in very peripheral vision. These results parallel Thorpe and colleagues' findings (Thorpe et al., 2001) which showed, that humans are very good at recognizing objects in the visual periphery (up to 70.5°). Likewise, Jebara and colleagues (2009) showed that participants recognized objects and faces above chance level up to 60°. Similar results have also been reported in the perception of low-level visual stimuli such as color (Naïli, Despretz, & Boucart, 2006). It has also been shown for biological motion stimuli, although there was always a disadvantage for recognition in the periphery in comparison to the visual abilities in central vision (Ikeda et al., 2005, 2013; B. Thompson et al., 2007). Our results extend those previous findings by demonstrating that we are even able to recognize complex stimuli like social actions in the visual periphery. Therefore peripheral vision might play a more important role in daily social interactions than just triggering gaze saccades towards conspicuous events in the periphery. One interesting observation in both experiments was the nonlinear decline of recognition performance with eccentricity. This seems to be at variance with previous studies that report a linear decline in the recognition of static objects (Jebara et al., 2009; Thorpe et al., 2001). To examine whether motion information is at the heart of the nonlinear decline, we compared the recognition performance of dynamic and static action stimuli in Experiment 2. The results showed a nonlinear decline of recognition performance with eccentricity for both types of stimuli. Therefore, the nonlinear decline of performance with eccentricity cannot be attributed to the presence of motion information in our stimuli. Jebara and colleagues (2009) used smaller pictures of faces, buildings and objects (10° visual angle) than in our study and found a linear decline while Thorpe and colleagues (2001) also report a linear decline although they used very large images (39° of visual angle high, 26° across) in which all displayed animals (e.g. an insect or a tiger) had more or less the same large visual size. Our stimuli are smaller than those of Thorpe and colleagues (2001) and larger than those of Jebara and colleagues (2009). Therefore, the size of the stimuli cannot be a factor explaining why performance declines differently for objects and social actions.

Our results are also relevant for the discussion about first and second levels in action recognition (see de la Rosa et al., 2015). One key feature of second level recognition is that it is faster than first level recognition (Rosch et al., 1976). While our results of Experiment 1 are in line with previous reports about action recognition being faster and more accurate at the second level (e.g. recognizing an action as a greeting), the results of Experiment 2 do not support this expectation. In Experiment 2, where all tasks have been equated in terms of response possibilities, we found the shortest reaction times and the highest recognition performance for the first level task, while second level recognition now seemed to be the more difficult task with longer reaction times and lower recognition performance. We argue that this reversal between Experiments 1 and 2 is not due to this equalization. If the equalization of response options was responsible for this response reversal, we would expect the pattern for first and second level recognition to reverse irrespective of the stimulus type (e.g. objects or social interactions). De la Rosa and colleagues (de la Rosa et al., 2011) as well as Grill-Spector and Kanwisher (Grill-Spector & Kanwisher, 2005) have shown that equating for response options in object recognition tasks does not change the pattern of results with regards to first and second level recognition.

What might be then the reason for the reversal of first and second level recognition between Experiment 1 and 2? We suggest that this reversal might be understood in terms of current action recognition models (Bertenthal & Pinto, 1994; Giese & Poggio, 2003; J. Lange & Lappe, 2006; Theusner et al., 2014). These theories assume that visual action information is mapped onto neuronal units that encode an action by means of many temporally ordered posture 'snapshots' (action template) that encode actions akin to individual frames of a movie showing a human action. We suggest that participants might have used this template matching mechanism more effectively in the first level task of Experiment 2 than in Experiment 1, thereby causing the reversal of the pattern. In particular, we asked participants to judge the target action in terms of a specific aspect (e.g. "Was the action a greeting: yes or no?" or "Was the action a handshake: yes or no?") in Experiment 2. Participants might have therefore benefited from top-down activation of the corresponding target action template, which resulted in matching all visual information onto this template in order to recognize the action. Such top-down influence is less efficient in the second level task than in the first level task. In the former case, visual information must be matched onto three (e.g. handshake, hugging and waving) instead of one action template in the latter case (e.g. handshake) in order to recognize the action. Hence, in the assumption that participants relied on top-down controlled template-matching strategy, one would expect first level recognition to be faster than second level recognition in Experiment 2. The same mechanisms could explain the recognition performance in Experiment 1 where no a-priori information about the target was provided. If participants relied on the same mechanism, they must have matched visual information against all six action templates in the first level recognition task. In the second level recognition task, participants could have chosen to monitor only one of the two levels (i.e. greeting or attack) and matched visual information onto the three corresponding action templates. In case there was no match, participants could have concluded that the non-chosen second level was displayed. In any case, matching visual information onto three action templates in the second level task should lead to better recognition performance than matching visual information onto six action templates in the first level recognition task. We found a decline in recognition performance that starts at smaller eccentricities in the first level task than in the second level and the valence tasks only in Experiment 1, but not in Experiment 2. This could be attributed to the fact that different action templates were needed to perform the first level task in Experiment 1. Participants needed more details to recognize the actions in order to categorize them. The visual resolution in the periphery was not sufficient for that task. A top-down controlled template-matching mechanism could therefore, in theory, explain the reversal of first and second level recognition performance between Experiment 1 and 2. Previous literature has already shown, with behavioral and neuroimaging evidence, the strong influence top-down mechanisms (e.g. attention, goal) have on the recognition of human actions (I. Bülthoff, Bülthoff, & Sinha, 1998; de la Rosa, Streuber, et al., 2014; Grezes, 1998; Hudson et al., 2015; J. Thompson & Parasuraman, 2012).

Overall, the classification of social actions in the different recognition levels seems to be less robust than for object recognition (de la Rosa et al., 2011; Grill-Spector & Kanwisher, 2005) in the sense that changing from a n-AFC task (with n>2) to a 2AFC task does not alter the overall pattern of results between the first and second level recognition tasks for object recognition.  Note that participants could have defaulted to the same second level recognition strategy in the valence recognition task. The reason for this is that the actions underlying the second level recognition (attack vs. greeting) were the same actions underlying the valence levels (negative vs. positive). Despite this possibility we find differences in performance between second level and valence recognition. This difference is suggestive of participants relying on at least partly different response strategies in these two tasks.

Furthermore, we would like to stress that second level and valence recognition in Experiments 1 and 2 were unlikely to be guided by a coarse assessment of the motion energy of the displayed action (e.g. more motion energy means negative or attack actions). When we changed the paradigm in Experiment 2 to make motion energy a much less effective cue for action classification by creating additional remapped distractor stimuli that had similar motion energy to the targets but did not show any meaningful actions, participants were still able to correctly classify the movies into their relevant categories.

To what degree could participants' performance relied on alternative recognition such as 'limb-spotting'? Because some actions are unique in the sense that they involved a unique limb, e.g. kicking action, it is possible that participants relied on monitoring the leg for the identification of the kicking action (i.e. defaulted to a 'limb spotting' instead of action recognition strategy). To address the issue, we examined the sensitivity results of each action in Experiment 2 (we did not look at RT since differences in RT might be simply owed to the length of the videos). A limb spotting strategy for kicking should lead to more accurate responses to this action in all tasks. Contrary to that expectation, kicking is an action that is recognized with an intermediate recognition performance in the basic and valence level tasks. As for the subordinate task, kicking is indeed associated with the best recognition performance. However, this recognition performance is not significantly different from another action, namely handshake ($t = 1.36$, $df = 31.17$, $p = 0.19$). Similarly, the results for hugging (the only action with bimanual movement) did not confirm the use of a spotting strategy. Therefore we think that limb spotting contributes little to the observed effects. Moreover, if participants would solely rely on limb spotting then recognizing actions carried out by the same limb (waving, slapping, punching) should be difficult to discriminate, which should lead to reduced recognition performance. However recognition performance of these actions is well above chance level.

Using a blocked instead of a random presentation of the different conditions might account for the high recognition performance in Experiment 2 to a certain degree. While we have no evidence on the effect of blocking vs. randomizing on recognition performance, we believe that this would have little influence on the results. We think that the nature of the task (i.e. looking out for an action in an array of action movies) in Experiment 2 led participants to rely on a top-down controlled recognition strategy in which participants monitor only the action channel relevant to the task in order to make the judgment whether the target action had been shown. It is well conceivable that participants can quickly switch between action channels that they would like to monitor. Hence if trials were randomized, participants were very likely to start monitor the channel corresponding to the target action by the time they have read the target action word presented at the beginning of the trial. As a result we would expect very little performance difference between blocked and randomized conditions. In our opinion, the largest performance change between blocked and randomized condition would be owed to a higher error rate because of the permanently switching action channels, which might lead participants to accidentally monitor the incorrect channel.

Last, we would like to point out our efforts to maintain a high ecological validity in our study for obtaining more robust results. While one might argue that stick-figures are not ideal, we think that in this initial study we found the simplest solution to avoid distracting discrepancies between body and face. Importantly, the large curved screen combined with the correction of distortion in the display ensured that actions were displayed at equal distance from the observer in a non-distorted ecological valid fashion. The use of life-size stimuli that were not scaled with eccentricity further allowed investigating the perception of actions in the visual periphery under more naturalistic conditions. In everyday life the size of another person does not change across eccentricities as long as the interpersonal distance does not change. Our study lines up with recent efforts that aim at investigating action recognition under more ecological valid viewing conditions, (e.g. Thornton, Wootton, & Pedmanson, 2014). These authors investigated the recognition of actions that were presented at various distances from the viewer and found that performance remains remarkably good even when the stimulus is moved far away along the line of sight. Here we also find high level of recognition despite lateral shifts of actions into the visual periphery.

## 2.6 Conclusion

The results of this study show that the recognition of another person's actions is well above chance level even in far periphery. In Experiment 1 Participants were able to categorize dynamic actions at the first and second level and recognized their emotional valence up to 60° eccentricity. In the second experiment we showed that the recognition performance decreased with eccentricity in a nonlinear fashion for static and dynamic actions. This indicates that the nonlinear decline is unlikely due to the motion information in the dynamic stimuli.

# 3 STUDY II: VIEWPOINT DEPENDENT ACTION RECOGNITION PROCESSES IN THE VISUAL PERIPHERY

Laura Fademrecht, Isabelle Bülthoff, Stephan de la Rosa

Max Planck Institute for Biological Cybernetics, Tübingen, Germany

## 3.1 Abstract

Recognizing actions of others in the periphery is required for fast and appropriate reactions to events in our environment. In this study we investigated the influence of the viewpoint of a social action on the recognition performance at fixation and in far visual periphery up to 75° eccentricity. Participants viewed a life-size stick figure avatar that carried out one of six motion-captured social actions (greeting actions: handshake, hugging, waving; attacking actions: slapping, punching and kicking). Participants either identified the actions as 'greeting' or as 'attack' or assessed the contained emotional valence. Reaction times were significantly faster for side views than for front views in both tasks. We argue that the side view (i.e. seeing the actor's profile) of an action might provide more visual information about the action as the front view (i.e. seeing the actor's front), which might help in peripheral vision where the visual resolution is highly decreased.

## 3.2 Introduction

For humans the recognition of another person's actions is crucial for fast and appropriate reactions to events in our environment. As social beings we usually encounter numerous different social actions in our daily life, that we usually recognize regardless of the viewpoint from which we see the action. The action 'waving' is recognizable whether we see it from the left, the right, the front or the back.

This process of recognition is closely related to the recognition of objects where the debate about viewpoint dependency is still ongoing. One theory (structural-description theory), concerning the recognition of objects shown from different viewpoints, argues that recognition performance is viewpoint-invariant within a range of viewpoints as long as all views show the same major component parts (geons) of the object and these parts' qualitative spatial relations (Biederman & Gerhardstein, 1993). Another theory (image-based theory) suggests a viewpoint-dependent mechanism and support a multiple-view approach, where objects are encoded as a set of view-specific representations that are matched to percepts using mental rotation or normalization procedures to transform the image to the closest known view (Tarr, 1995). Tarr and Bülthoff (1998) then proposed a model that combines the most appealing aspects of image-based and structural-description theories for visual object recognition to explain findings from psychophysics, neurophysiology and machine vision (see also Foster and Gilson 2002).

Image-based and structural-description theories have been concerned with the recognition of unfamiliar objects that have been learned only from certain viewpoints. However, another person's actions, we encounter everywhere in our environment in daily life. Therefore, our visual system is already trained to seeing social actions from various different viewpoints and also in various different manners. Social action stimuli are nowhere unfamiliar to the human visual system. Nevertheless, there is some evidence that the recognition of human actions is indeed viewpoint dependent. Jokisch and colleagues (2006; and Troje et al., 2005) showed that the recognition performance of one's own walking pattern was viewpoint-independent whereas the recognition performance for other familiar individuals was better for frontal and half-profile view than for profile view. Daems and Verfaillie (1999) showed that priming stimuli that had the same view as the test stimuli were more effective than mirror-image prime stimuli. There is some physiological evidence that the recognition of human bodies is viewpoint-dependent as well. Perrett et al. (1989) localized cells in the temporal cortex through single-cell recording that are only activated when seeing faces or bodies from a particular viewpoint. Verfaillie (1993) examined the effects of depth rotation using short-term priming with point-light walkers and were able to show that priming effects only occurred when the priming walker and the test walker had the same orientation.

The apparent existence of viewpoint-dependency of action recognition even though the stimuli are not completely unfamiliar could be due to the fact that in case of action recognition the issue of viewpoint-dependency enters a new dimension, since actions are moving stimuli and can therefore take an infinitive number of different views that fall as 2D projections on the retina (Daems & Verfaillie, 1999).

Another aspect that might play a role in the viewpoint-dependency of the recognition of social actions is the social component of these specific types of actions. Since social actions might elicit some emotional response in the viewer it could be a crucial difference seeing a social action that is directed to oneself (front view) in contrast to an action that is directed elsewhere (side view) where the viewer remains an impartial observer of the action. Assuming this would imply that social perception is fundamentally different in scenarios of social interaction, compared to a scenario of mere detached observation of a social action (Schilbach et al., 2013). However, it has remained unclear whether neural processes are influenced by the degree to which a person feels involved in an interaction (Schilbach, 2010). Results of Schilbach et al. (2006), demonstrate that participants give socially relevant facial expressions a higher rating when they are directed toward the participants and that neural activation patterns differed when the facial expressions were directed towards the observer. Nevertheless, it remains unclear to what degree a person feels immersed in interaction when they merely see human movement that is either directed toward them or directed somewhere else.

Considering daily life scenarios however, we must be aware that actions do not only appear in foveal vision or at the point of fixation, but can appear anywhere in our visual field. Although the visual abilities diminish towards the periphery people are still able to recognize actions that appear in far visual periphery with a high accuracy (Fademrecht, Bülthoff, & de la Rosa, 2016). In the visual periphery the influence of the emotional content of social actions might even be more important. Findings of Holmes et al. (2005) suggest that emotional information can be readily extracted from low spatial frequency input in the visual periphery, activating magnocellular pathways and the amygdala. On the one hand this might influence the degree to which the person feels engaged in interaction. On the other hand it is not clear whether a person really feels engaged in an interaction that is in fact directed towards them but is viewed from the corner of the eye. In line with this we used two different recognition task. In one recognition task we asked for the semantic basic level of the action (de la Rosa, Choudhery, et al., 2014), meaning that participants reported whether they saw a greeting or an attack. In the other task participants were asked to report the emotional valence of the action (valence task). In a previous study (Fademrecht et al., 2016), we were able to show a difference in recognition performance between the two tasks. Evaluating the emotional valence of the action could elicit a stronger feeling of social involvement than the semantic categorization of an action and might lead to a stronger effect of the directedness of the observed action.

The focus of this article is to investigate the viewpoint-dependency of social actions that are shown at fixation and in the peripheral visual field.

## 3.3 Methods

*Participants:* 30 participants were recruited (11 males, 20 females) from the university community of Tübingen. The age ranged from 21 to 32 years (mean: 25.5). All participants received monetary compensation for their participation and gave their informed written consent prior to the experiment. The participants had normal vision or corrected their visual acuity using contact lenses. The study was conducted in line with Max Planck Society policy and has been approved by the Max Planck ethics committee.

*Stimuli:*      The six actions were recorded via motion capture using Moven Suits (XSens, Netherlands). The Xsens MVN Suits consists of 17 inertial and magnetic sensor modules, which are placed in an elastic lycra suit worn by the actor performing the actions that were recorded. The sampling rate was 120 Hz. Three actions with positive emotional valence (handshake, hugging and waving) and three actions with negative valence (slapping, punching and kicking) were acted out by six different actors (three male, three female). Every action was repeated six times by each actor, leading to 216 stimuli in total. The actions lasted between 800 and 1500 ms and each action started with the actor standing in a neutral position (N-pose) and ended with the subjective peak frame of the action. The subjective peak frame of an action is considered to be the point in time just before the actor started moving back into the neutral position.

The motion data was mapped onto a grey life-size 'stick figure avatar' that was projected on the screen (height: 25.15° visual angle). The figures, were either oriented towards the participant or orthogonal to the participant's direction of view. A stick figure was used instead of a full-fleshed avatar in order to prevent any other visual cues like appearance or gender from influencing participant's recognition judgements. Furthermore, the use of this simple figure avoids discrepancies between a moving realistic avatar and his face that would have been static, thus exhibiting a fixed neutral facial expression with a fixed gaze. Second, thereby we also avoided the discrepancy of having an avatar with a fixed neutral expression performing actions with an emotional valence. We refrained from using a point light display because it might give too little visual information for the lower resolution of the visual system in far periphery.

*Apparatus:*      Stimuli were presented on a large panoramic screen with a semi-cylindrical projection system. The semi-circular wide screen was 7 m long (diameter) and 3.2 m in high (230° horizontally, 125° vertically). Six EYEVIS LED DLP projectors (1920x1200, 60Hz) were used to display the stimuli against a grey background. The geometry of the screen can be described as a quarter-sphere. The visual distortions caused by the curved projection screen were compensated virtually with the use of warping technology software. With this screen visual stimuli can be presented to the whole horizontal human visual field. Participants placed their head on a chin and forehead rest. An eye tracker (Eyelink II, SR Research Ltd., Canada) was used to control for eye movements. If the participant's gaze shifted more than 2° away from the fixation cross in the middle of the screen (0°) the trial was discarded. When the stick figure was presented at 0° it was presented behind the cross. The Unity 3D (Unity Technologies, USA) game engine in combination with a custom written control script was used to control the presentation of the stimuli and to collect responses.

*Procedure and Design:* The experiment started with the explanation of the following experimental procedure (Fig). Each trial began with the presentation of a fixation cross in the middle of the panoramic screen and the eye tracker started to record the eye movements. Participants were instructed to keep their gaze fixated on the fixation cross, while the stick figure appeared at one of the eleven positions in the participant's visual field. Trials with a gaze shift larger than 2° were discarded from the analysis (0.6% of the trials). The task was to answer one of the following two questions in a between subject design. Participants either answered the question "Was the action a greeting or an attack?" meaning that they categorized the action on a basic level (basic level task), or they answered the question "Was the action positive or negative?" to evaluate the emotional valence of the viewed action (valence task). Participants were instructed to answer as fast and accurately as possible. They could give the answer as soon as the stick figure appeared on the screen. In case participants did not respond before the end of the animation sequence, a prompt appeared on the screen, displaying the question and the pre-defined response keys on a keyboard (1 or 0 on the keyboard). Three of the actions had a negative emotional valence (kicking, punching, slapping) and three had a positive emotional valence (handshake, hugging, waving). Each of the six actions was presented 120 times. We manipulated the position of the stick figure in the participant's visual field (eccentricity), so that it appeared 0°, 15°, 30°, 45°, 60° or 75° away from fixation, randomly either on the left or the right side of the screen (fig). In half of the trials the action was oriented towards the participant and in the other half of the trials the actions were directed orthogonal to the participant's view. The emotional valence, the orientation of the actions and the positions on the screen were counterbalanced within participants. That is, each action was presented at each position 20 times in both orientations, which resulted in a total of 480 trials (=20 repetitions x 6 positions x 2 emotional valences x 2 orientations). The 216 stimuli were shown 2.2 times in the 480 trials. Actions, positions and orientations were random but counterbalanced within one participant.

The two recognition tasks were performed by two separate groups of 15 participants, hence recognition task was a between subject factor. Position and orientation were within subject factors. Reaction times and accuracy served as dependent variables. In the valence recognition task participants had the answer options "positive" and "negative", while in the basic level task the answer options were either "greeting" or "attack". At the beginning of an experiment participants received a short training of 10 trials in order to get familiarized with the setup and the task. The stimuli used in the training trials were different from stimuli in the test trials.

## 3.4 Results

Accuracy and reaction times are considered to be measurements of recognition performance. An influence of the stimulus viewpoint on either one of these measurements can therefore be considered as a change in recognition ability due to the manipulation. Accuracy and reaction times are presented separately.

**Reaction times**

Participant's reaction times increase nonlinearly with eccentricity for both recognition tasks and both stimulus orientations (Figure 2). A three-way repeated measures ANOVA revealed a significant main effect of eccentricity ($F(5, 140) = 112.56$, $\eta^2_p = 0.36$, $p < .001$) and therefore supports the visual impression of increasing reaction times with increasing eccentricity. The ANOVA showed no other main effects (all $p$ values higher than .05), respectively there was no effect of the recognition task on participant's reaction times. There was a significant interaction of eccentricity and stimulus orientation ($F(5, 140) = 8.49$, $\eta^2_p = 0.003$, $p < .001$), which indicated that the increase of reaction times with eccentricity was different for the two stimulus orientations. A pairwise comparison via t-test showed, after Bonferroni correction (significance level: $p = .008$), no significant difference for the two stimulus orientations at 0° ($t_{paired} = 3.12$, $df$

= 29, $p$ = .92) and 15° ($t_{paired}$ = -2.13, $df$ = 29, $p$ = .04). For all other more peripheral presentations there was a significant difference between the two stimulus orientations (30°: $t_{paired}$ = 3.12, $df$ = 29, $p$ = .004; 45°: $t_{paired}$ = 3.12, $df$ = 29, $p$ = .004; 60°: $t_{paired}$ = 5.24, $df$ = 29, $p$ < .001; 75°: $t_{paired}$ = 4.12, $df$ = 29, $p$ < .001). The side view of the stimulus led to shorter reaction times than the front view beyond 15° eccentricity. The ANOVA showed further that all other two-way interactions and the three-way interaction were non-significant (all $p$ values higher than .05).

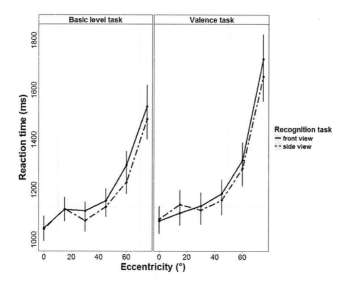

*Figure 2: Means and standard errors of reaction times (ms) as a function of eccentricity for front and side view in the two recognition tasks.*

**Accuracy**

The accuracy for both tasks and the two stimulus orientations decreased nonlinearly with eccentricity and was above chance level up to 75° (Figure 3). The results of a three-way repeated measures ANOVA showed a significant main effect of eccentricity ($F(5, 140)$ = 1.46, $\eta^2_p$ = 0.69, $p$ < .001), which indicates a significant decrease of accuracy with eccentricity. There was no significant main effect of recognition task ($F(1, 28)$ = 2.9, $\eta^2_p$ = 0.03, $p$ < .09) and stimulus orientation ($F(1, 28)$ = 1.17, $\eta^2_p$ = 0.002, $p$ = .29). The two-way interaction between eccentricity and stimulus orientation was significant ($F(5, 140)$ = 4.51, $\eta^2_p$ = 0.03, $p < .001$). This shows that the there is a difference between the two stimulus orientations but only for certain eccentricities. In fact, after Bonferroni correction (significance level: $p = .008$), only at 45° eccentricity there was a significant difference between front view and side view ($t_{paired}$ = 3.14, $df$ = 29, $p = .003$). The front view led to a higher accuracy only at 45°. At all other eccentricities, there was no significant difference between the two stimulus orientations (all $p$ values higher than the significance level after Bonferroni correction). All other two-way interactions and the three-way interaction were non-significant (all $p$ values higher than .05).

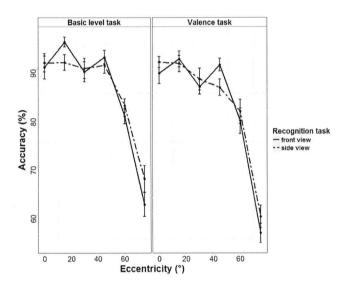

*Figure 3: Means and standard errors of accuracy (%) as a function of eccentricity for front and side view in the two recognition tasks.*

## 3.5 Discussion

Our results showed nonlinearly decreasing recognition performance for the action stimuli with eccentricity. The accuracy was above chance level for all tested eccentricities (up to 75°). The recognition task had neither an effect on participant's reaction times, nor on the accuracy. Overall the results reveal an extremely high performance for the recognition of social actions even at such far eccentricities as 75°.

Thorpe and colleagues (2001) were already able to show a performance above chance level for object recognition at 70.5°. We show with this study that a similarly high performance can be reached for the recognition of social actions up to 75°.

The major effect of stimulus orientation was visible in the reaction time results. We found shorter reaction times for the side view of the actions in far periphery, (beyond 15° eccentricity), indicating that in far periphery people are faster when they view an action in the side view. This could be explained by the fact, that the amount of visual information about the stimulus is higher in the side view. Seeing a punch from the front view, where there is just a fist coming towards the observer, and the side view, where the stretching of the whole body is visible, could help the recognition in far periphery.

The stimuli used in this study are social actions, therefore it is important to also consider the aspect of social interaction in our experiment. Schilbach and colleagues, who argue towards a second-person neuroscience, claim that the perception of self-directed stimuli brain regions that have been related to emotional and evaluative processing (Schilbach et al., 2006). Perceiving a social stimulus that is directed towards the observer (here front view) are perceived from a second-person perspective. Here the observer is part of a social interaction and not only a pure observer of someone else's actions. Seeing an action that is directed somewhere else than the observer (here side view) can be considered as perception from the third-person perspective since the observer is not part of the interaction and is merely an impartial viewer. Being part of a social interaction might trigger different processes than those involved in observing someone else interact (Schilbach et al. 2013). Based on this, one might come to the conclusion that the recognition of social actions from the second-person perspective (front view) should have an advantage over the recognition of an action from the third-person perspective (side view). In our results we find a significant difference between the two perspectives. However, in our experiment, this difference we found only in far periphery but not in central vision and near periphery. In far periphery the third-person perspective (side view) led to shorter reaction times over the second-person perspective (front view). Therefore our results contradict the prediction of an advantage of second-person over third person perspective. But we need to take into account that in peripheral vision the lower resolution and therefore the need of more visual information could overwrite the effect of the viewer's perspective.

.

# 4 STUDY III: MEASURING PERCEPTIVE FIELD SIZES OF ACTION SENSITIVE PERCEPTUAL CHANNELS

Laura Fademrecht [1], Nick E. Barraclough [2], Isabelle Bülthoff [1], Stephan de la Rosa [1]

[1] Max Planck Institute for Biological Cybernetics, Tübingen, Germany

[2] University of York, York, UK

## 4.1 Abstract

Humans are social beings who interact with other people. For successful interaction, one needs to be able to recognize the actions of another person. Action recognition is an important part of our daily life and happens not only in our central visual field. In a recent study (Fademrecht et al., 2016), we found that peripheral recognition of actions is surprisingly good even in far periphery up to 60° eccentricity. Moreover we observed a nonlinear decline of recognition performance in the periphery. These results are surprising in the light that visual object and face recognition typically linearly declines with eccentricity – a result attributed to the coarser spatial sampling of visual information in the visual periphery. Here we examined the spatial sampling area of action sensitive mechanisms behaviorally (perceptive field sizes) by means of action adaptation. Participants were adapted to a handshake or a punch action at one position in the visual field and tested with an ambiguous morph between the two actions at this and several other position within the visual field. Action adaptation aftereffects where largest at the adapted location and decreased with increasing distance from the adapted location. Interestingly this decline with spatial distance was stronger in the periphery than in the fovea suggesting the perceptual field sizes decrease with increasing eccentricity.

## 4.2 Introduction

Most of the visual information that impinges on the retina falls into the visual periphery and not in the fovea. Yet the majority of studies examining visual recognition focus their interest on the fovea. One obvious explanation for this is that visual recognition of objects, faces, actions, and letters is best in the fovea. As stimuli are presented more peripherally, recognition performance declines linearly with eccentricity. This decline of recognition performance in the periphery has been often explained in two ways. Firstly, in terms of the spacing of the photoreceptors in the retina and of ganglion cells (Anderson et al., 1991; Banks, Sekuler, & Anderson, 1991; Ennis & Johnson, 2002; Frisen & Glansholm, 1975; Popovic & Sjöstrand, 2001, 2005) and secondly, in terms of increasing receptive field sizes towards the periphery and the concomitant reduced cortical representation of the peripheral visual field (Daniel & Whitteridge, 1961). Both, scatter and size of receptive fields of ganglion cells are small in the foveal region and large in the visual periphery (David H Hubel & Wiesel, 1974). The receptive field size increases linearly, with a corresponding decreasing visual resolution in the visual periphery (D H Hubel & Wiesel, 1965; Wilson & Sherman, 1976). The linear decline of visual recognition performance has been shown for low-level visual stimuli (Hansen, Pracejus, & Gegenfurtner, 2009; Jacobs, 1979; Tynan & Sekuler, 1982) and also for objects and scenes (Jebara et al., 2009; Nelson & Loftus, 1980; Thorpe et al., 2001).

We recently observed a notable exception to the linear decline of recognition performance in the visual periphery (Fademrecht et al., 2016). We found the recognition of life-size human actions to be remarkably good in far periphery up to 60° eccentricity (Fademrecht et al. 2016). In particular, the decline of action recognition performance was nonlinear with no statically significant decrease compared to central vision up to 30° eccentricity. It was best described by a cubic trend.

Here we were interested why action recognition performance changes very little in the near periphery despite the loss of visual resolution caused by larger receptive fields. We therefore decided to measure the behavioral correlates of receptive fields, namely perceptive fields. A perceptive field is considered to be the psychophysical correlate of the physiologically determined receptive field (Neri & Levi, 2006; Spillmann, 2014; Troscianko, 1982). In their reviews, Neri and Levi (2006) as well as Spillmann (2014) provide an overview over experiments in psychophysics and physiology and show a remarkable commonality between perceptive fields measured by means of behavioral experiments and receptive fields measured using single-cell recordings. Receptive fields have been shown to increase in size with increasing eccentricity (D H Hubel & Wiesel, 1965; Wilson & Sherman, 1976) and to scale along the hierarchy of the visual pathways in the brain (Freeman & Simoncelli, 2011). Neurons in higher processing areas in the brain pool information from multiple cells of the lower processing stages (Freeman & Simoncelli, 2011; Giese & Poggio, 2003), leading to an increase in receptive field size. The extent of perceptive fields has mainly been assessed for low-level visual stimuli (Neri & Levi, 2006; Ransom-Hogg & Spillmann, 1980; Troscianko, 1982), showing an increase of perceptive field sizes with increasing eccentricity similar to receptive field sizes. Human actions are complex high-level visual stimuli for which one can assume large receptive and perceptive fields. It is widely believed, that perceptive fields for the recognition of human actions are widely spread, allowing position insensitivity of action recognition to a large degree (see for example Giese and Poggio 2003).

In the current study, we aimed to measure the perceptive field size underlying action sensitive visual processes. A method optimal for selectively targeting visual processes is visual adaptation. In an adaptation experiment participant are exposed to a visual stimulus (adaptor) for a prolonged amount of time. Typically it is found that this prolonged exposure to the adaptor transiently changes the subsequent percept of an ambiguous test stimulus. For example after adapting to a red square a white square is perceived with a greenish tint (adaptation aftereffect). Adaptation effects are typically explained in terms of a response change visual processes that are shared between adaptor and test stimuli (e.g. response change of the red channel which influences the pooled response across several colour channels) (M. A. Webster, 2011). These action adaptation effects are most well-known for low level visual features such as colour

(McCollough, 1965) and orientation (C.W.G. Clifford, 2002; Colin W G Clifford, Wyatt, Arnold, Smith, & Wenderoth, 2001) but have also reported for more complex visual patterns such as motion (G. Mather, Verstraten, & Anstis, 1998), and faces (Kovacs, 2005; Rhodes et al., 2003; M. a Webster & MacLeod, 2011), and actions (Barraclough et al., 2009; de la Rosa et al., 2016; de la Rosa, Streuber, et al., 2014). As for action recognition, we have recently introduced an action adaptation paradigm that is able to target visual processes underlying the human ability to categorize actions by means of action morphing. Here we used action adaptation to assess the perceptual field size of visual processes sensitive underlying action categorization.

In order to measure the perceptual field size of action categorization processes, we visually adapted participants at one location within their visual field. We afterwards probed the adaptation aftereffect at several other locations within the visual field including the adapted location. We reasoned that adaptation most effectively changes the response of action recognition processes which have their perceptual field at the adaptation location. Correspondingly we expected the adaptation to be largest when the test stimulus is presented at the location of adaptation. If we move the test stimulus away from the perceptual field centre (i.e. the location of adaptation) into its periphery, the test stimulus should elicit a reduced response in the corresponding action recognition process. Consequently we expect a reduced adaptation transfer from the adapted to the probed location, which should result in a smaller adaptation effect. Finally, if we were to move the test stimulus outside of the adapted perceptual field, the action recognition process should not respond to the stimulus anymore. Consequently we would not expect an adaptation transfer between the adapted and probed location and hence there should be no adaptation effect. In essence, the magnitude of the adaptation effect as a function of spatial separation between adaptor and test stimulus should give a rough estimate for the perceptual field size at the adapted location.

To this end participants saw a life sized human-like avatar that carried out an action (handshake or punch) for a prolonged amount of time at one position in their visual field and were subsequently tested with an ambiguous test stimulus somewhere in their visual field (including the adapted position). We measured the magnitude of the adaptation effect as a function of spatial adaptor-test separation.

Finally, we were interested in whether the perceptive field size is able to explain our previous experimental results. For this reason we examined the relation between perceptual field sizes and action recognition.

## 4.3 Methods

*Participants:*   45 subjects (20 males, 25 females) from the local community of Tübingen participated in the experiment. All participants received monetary compensation for their participation. Their age ranged from 19 to 37 years (mean: 25.3). The visual acuity of the participants was either normal or corrected to normal, using contact lenses. Participants gave their informed written consent form prior to the experiment. The study was conducted in line with Max Planck Society policy and has been approved by the University of Tübingen ethics committee.

*Apparatu*s:   For stimulus presentation, a large panoramic screen with a half-cylindrical virtual reality projection system was used (Figure 1). The almost semi-circular screen is 3.2 m high and 7 m long. It covers 230° horizontally and 125° vertically of the visual field of our participants seated in front of the screen. The format of the screen can be described as a quarter-sphere. Six LED DLP projectors (1920x1200, 60Hz; EYEVIS, Germany) were used to display the stimuli against a grey background on the screen. We used warping technology software (NVIDIA, Germany) to compensate for the visual distortions of the display caused by the curved projection screen. All stimuli were presented at a virtual distance of 3 m. Participants sat on a chair in front of a desk. The desk was placed in the middle of the screen arena. Participants placed their head on a chin and forehead rest, mounted on the desk. During each experimental trial, they were required to keep their eyes focused on a white fixation cross, placed on the screen straight ahead of them.  This position of the cross was defined as 0° position. An eye tracker (Eyelink II, SR Research Ltd., Canada) was used to control for their eye movements. When the stick figure was presented at 0°, it was presented behind the cross. The Unity 3D (Unity Technologies, USA) game engine in combination with a custom written control script was used to control the presentation of the stimuli and to collect responses given by the participants via predefined keys on a keyboard.

*Stimuli:* A handshake and a punch action performed by one actor were recorded via motion capture using Moven Suits (XSens, Netherlands). The Xsens MVN Suit consists of 17 inertial and magnetic sensor modules, placed in an elastic lycra full-body suit worn by the actor. The sampling rate was 120 Hz. Both actions started with the actor standing in a neutral position and lasted 708 ms. Each action ended at the peak frame of the action, which was specified as the point in time just before the actor started moving back to the neutral position.

The motion data was mapped onto a grey 'stick figure avatar', in order to prevent visual cues like appearance and gender to influence participants' decisions. Using a stick figure had the advantage that we did not have to record facial movement information (e.g. expression and gaze) and hand and foot motions. We favoured the use of a stick figure over a dynamic point-light display because the sparse structure of the latter might unduly hinder recognition because of the decreasing spatial resolution of the visual system towards the periphery. The avatar was life-size, with a height of 170 cm, (subtending approximately 25° of vertical visual angle). The position midway between both hips of the stick figure avatar defined the general position of the stick figure avatar in the visual field (and on the screen). The stimuli were oriented toward the participant and presented along the same latitude (i.e. on the same horizontal axis) at any position on the screen.

We created adaptor and test stimuli derived from the action sequences. For the test stimuli, we used a morphing algorithm that allowed creating body motions in between the punch and the handshake action  Various weighted averages of the positions in space of each joint on the body (for example the elbow) in the two action sequences were calculated for each action frame. The point of subjective equality was defined as the weighted average (morph ratio) at which the whole action looked completely ambiguous. We used the following seven morph levels, which we determined to elicit an ambiguous percept of the actions in six participants in a pilot study: 36%, 38%, 40%, 43%, 45%, 48%, 50%, 53%, and 55% of the punch action. The 100% punch and 100% handshake actions served as adaptor stimuli.

*Procedure and Design:* Participants sat in the middle of the screen arena (as indicated by a fixation cross) with their heads rested on a chin rest. A fixation cross in the centre of the screen was present on the screen during the whole duration of each stimulus presentation. To control for eye movements we mounted an Eyelink II eye tracker on the chin rest and recorded participants' eye movements (trials with a gaze shift larger than 2° were discarded from all analyses; less than 1% of the trials). Participants were instructed to fixate the cross and informed that trials in which their gaze moved away from the cross would be excluded.

First, we probed participant's perception of the test stimuli without visual adaptation (baseline condition). In the baseline condition, each trial started with the presentation of the fixation cross and, after 500 ms, the test stimulus was presented for 708 ms. The question "What did it look more like?" and the response options "handshake" and "punch" along with their respective answer keys (0 and 1) appeared on the screen 500 ms after the test stimulus was presented. Participants responded using corresponding keys on a keyboard. The seven test stimuli were presented three times at the seven positions in the visual field (-60°, -40°, -20°, 0°, 20°, 40°, 60°) in random order.

Next, participants continued with the experimental conditions. In one condition the handshake action was used as adaptor, in the other the punch action served as adaptor. During all experimental conditions participants were asked to look at the fixation cross in the middle of the screen. At the beginning of each condition, participants watched the adaptor 26 times before the actual experimental trials started. An experimental trial consisted of four adaptation presentations (each 708 ms; Adaptor ISI: 500 ms) which was followed by a beep sound (frequency: 1000 Hz). The beep sound always preceded the presentation of the test stimulus and was meant to help participants which stimulus to judge (i.e. the one after the beep). The test stimulus was presented for 708 ms. Subsequently, the question "What did it look more like?" and the response options "handshake" and "punch" appeared on the screen 500 ms after the test stimulus was presented. Participants responded using corresponding keys on a keyboard. Participants were instructed attend to the adaptor during the adaptation phase and to decide whether the test action looked more like a handshake or a punch.

Three separate groups of 15 participants were tested with a different position for the adaptor (0°, -20° or -40°). Hence, adaptor position was a between-subjects factor. For each group we tested two experimental conditions. Each experimental condition probed only one adaptor (handshake vs. punch) at the same adaptor position in the visual field. The testing order of the two experimental conditions was counterbalanced across participants. Within each experimental condition the test stimulus varied its position with every trial randomly (for adaptation at 0°: -60°, -40°, -20°, 0°, 20°, 40°, 60°; for adaptation at -20° (left of the fixation cross) and -40° the positions -70° and 70° were added to the test position at 0°). Each of the test stimuli was presented three times at each of the seven positions (respectively nine positions for adaptation at -20° and -40°) in the visual field. Participants were instructed to report their subjective feeling, without trying to be constant in their answer patterns. Responses were given via key press on a keyboard. The total duration of an experimental condition was about 50 minutes.

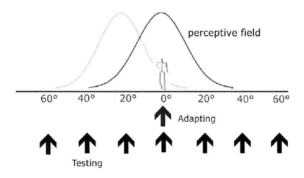

Figure 4: Experimental procedure

## 4.4 Results

The adaptation effect was calculated as the difference in proportion punch responses between the handshake and punch adaptation conditions.

(orange). For all three adaptation positions the adaptation effect is highest at the position of adaptation (adaptation at 0°: 0.57; adaptation at -20°: 0.49; adaptation at -40°: 0.29), and decreases as adaptor-test distance increases. For adaptation at 0° eccentricity we find, using Bonferroni corrected t-tests, a significant adaptation effect for testing at 0° ($t_{paired}$ = 8.88, $df$ = 14, $p < .001$), at -20° ($t_{paired}$ = 6.59, $df$ = 14, $p < .001$), at 20° ($t_{paired}$ = 4.79, $df$ = 14, $p < .001$) and 40° ($t_{paired}$ = 3.06, $df$ = 14, $p < .01$) eccentricity. At all other testing positions there was no significant adaptation effect (all $p$ values higher than .01, the significance level after Bonferroni correction). For adaptation at -20° we identified significant adaptation effects only for testing at -20° ($t_{paired}$ = 6.1, $df$ = 14, $p < .001$). For all other testing positions the $p$ value was higher than 0.0125 (significance level after Bonferroni correction). When we adapted at -40° we find the only significant adaptation effect for testing at -40° eccentricity ($t_{paired}$ = 4.4, $df$ = 14, $p < .001$), while for all other testing positions the adaptation effect was nonsignificant (all $p$ values were higher than 0.0125, the significance level after Bonferroni correction).

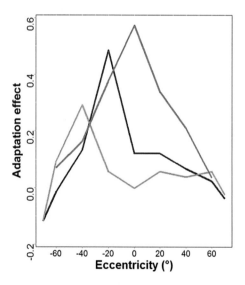

Figure 5: Distribution of the adaptation effect over the testing positions for adaptation at 0° eccentricity (cyan), adaptation at -20° eccentricity (blue) and adaptation at -40° eccentricity (orange)

Receptive fields for a population of neurons have been approximated with a two-dimensional Gaussian function (Amano, Wandell, & Dumoulin, 2009; Dumoulin & Wandell, 2008; Zuiderbaan, Harvey, & Dumoulin, 2012). Since it is reasonable to assume that the adaptation paradigm probes a population of neural responses here along only one (i.e. the horizontal) dimension, we approximated the shape of the perceptive fields with a one-dimensional Gaussian functions (equation 1). The fits were carried out by means of the 'gfit' function in MATLAB. We fitted the following function:

$$f(x) = m \cdot \left( \frac{1}{\sqrt{2 \cdot \pi \cdot \sigma^2}} \right) \cdot e^{-1 \cdot \left( \frac{(x-\mu)^2}{2 \cdot \sigma^2} \right)} \qquad \text{Equation 1}$$

The parameter 'μ' indicates the position of the maximum of the Gaussian function and therefore, in this case, the position of adaptation. 'm' is a scaling parameter that scales the function along the y axis. The Gaussian functions fit the data well with a mean $R^2$ of 0.98.

We defined the *Full Width at Half Maximum* (FWHM) of the Gaussian function as spatial extent of the perceptive field. For adaptation at 0° eccentricity, we found a spatial perceptive field of 62.74°, whereas for adaptation at peripheral positions we find smaller perceptive fields (for -20°: 29.06; for -40°: 25.72°). The Gaussian functions of the different adaptation positions overlap to a large degree, indicating that the channel at 0° influences also the recognition of actions at adjacent positions.

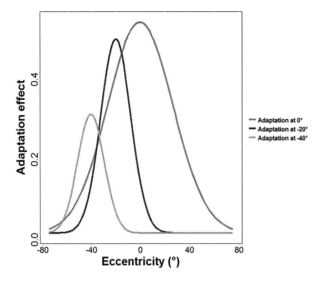

*Figure 6: Gaussian Functions fitted to the data of the distribution of the adaptation effect over the testing positions for adaptation at 0° eccentricity (cyan), adaptation at -20° eccentricity (blue) and adaptation at -40° eccentricity (orange)*

**The relationship between perceptive field size and action recognition**

Can the perceptive fields measured in this study explain action recognition performance found in the previous study (Fademrecht et al 2016)? For this reason we examined the relationship between the spatial extent of action sensitive channels and the recognition performance for social actions from a previous experiment (Fademrecht et al., 2016). Typically, the summed output of perceptual channels and human performance correlate well for low level and more complex visual stimuli (M P S To, Baddeley, Troscianko, & Tolhurst, 2011; M. To, Lovell, Troscianko, & Tolhurst, 2008; Michelle P. S. To, Lovell, Troscianko, & Tolhurst, 2010).

For this reason we tried to establish a mathematical relationship between the spatial extent of perceptual fields of action sensitive channels and the measured recognition performance. A summation of the Gaussian functions with a scaling factor 'w' and an offset 'c' represent a good fit of the recognition performance with an $R^2$ of 0.99:

Equation 1:

$$d'(eccentricity) = c + w \cdot \sum_{i}^{n=3} G_i(eccentricity)$$

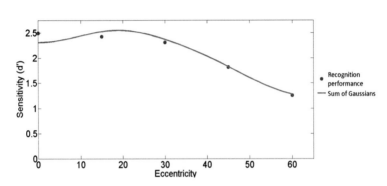

*Figure 7: Sum of Gaussians (grey) fitted to the recognition performance data (green) of a previous experiment (Fademrecht et al., 2016)*

Therefore showing that the model, based on the distribution of perceptual fields of action sensitive channels, can predict the recognition performance for social actions that appear somewhere in the visual field.

## 4.5 Discussion

In the current study, we used an action adaptation paradigm to investigate the size of perceptive fields of action sensitive channels. We then examined whether perceptual fields and recognition performance are related in order to see whether perceptual field size is a potential candidate for explaining action recognition in the periphery.

We found several interesting results. First, for all adapted locations, the adaptation effect is largest at the adapted position. Second, the magnitude of the adaptation effect at the adapted position decreased with increasing eccentricity. Third, FWHM decreases with increasing eccentricity. We discuss the implication of each

This is interesting for the discussion about the origin of the action adaptation aftereffect. Here an open question remained to what degree high-level (e.g. decision) mechanisms mediate the action adaptation aftereffect. Cognitive factors his indicates that the action adaptation aftereffect is a locally bound effect.

Our results show that the adaptation effect not only occurred at the position of adaptation but at adjacent positions as well. Although, the further away from the adaptation position the test stimulus was shown, the smaller was the adaptation effect. Therefore, the adaptation effect can be used as a tool to measure the spatial extent of perceptive fields of neuronal populations.

For adaptation at fixation, we found the largest perceptive field of action sensitive channels. Toward the periphery, the spatial extent of the perceptive fields decreased, as well as the amount of activation of the channel. We therefore find that the perceptive fields of neuronal populations seem to decrease with eccentricity, whereas receptive field sizes of ganglion cells increase with eccentricity. One possible explanation could be the overall lower visual abilities in the periphery. The smaller perceptive field could be the result of a lower recognition performance and therefore a lower adaptation effect at peripheral positions. In a previous experiment on the other hand, we were able to show that the recognition performance for social actions is similar to the performance in central vision between -45° and 45° eccentricity (Fademrecht et al., 2016). This would indicate that the recognition performance at -20° does not differ significantly from the recognition performance in central vision and thus should not result in such a large difference in measured perceptive fields. More research is needed to further investigate the decreasing receptive field sizes of action sensitive neuronal populations with eccentricity.

Since the perceptive fields overlap to a large degree and the action channel in central vision influences the recognition of actions at adjacent positions, the recognition performance is high for positions with a high influence of the action sensitive channel in central vision. Only for eccentric positions with low or no influence of the central channel, we find a decrease in action recognition performance. This information therefore provides an explanation for the surprisingly high action recognition performance in the visual periphery.

Interestingly we were able to use the knowledge we gained about perceptual fields of action channels to model the action recognition performance in a previous experiment (Fademrecht et al., 2016). By summation of the Gaussian functions that represent the spatial extent and the activation of action sensitive perceptual channels, we could model the action recognition performance throughout the visual field. This model can give us an explanation for the high and nonlinearly declining recognition performance, that we found previously (Fademrecht et al., 2016). Since the perceptual fields overlap to a large degree and the action channel in central vision influence also the recognition of actions at adjacent positions, the recognition performance is high for positions with a high influence of the action sensitive channel in central vision. Only for eccentric positions with a small or no influence of the central channel, we find a decrease in action recognition performance. This information therefore provides an explanation for the surprisingly high action recognition performance in the visual periphery.

Our results are also important considering position sensitivity of action recognition processes. Increasing position invariance is assumed for human actions with increasing cortical hierarchy, although the extent of this position insensitivity remains still unclear (Fleischer et al. 2013). The recognition of human actions usually takes place not in a fixed portion of the visual field because actions are moving and can include different parts of the visual field over the course of time. Therefore, position invariance to a large degree facilitates the recognition of another person's actions. However, complete position invariance in higher stages of processing would on the other hand hinder the recognition in cases where for example other actors or objects are involved in an observed scene. Furthermore, action recognition requires a good balance between selectivity and generalizability. On the one hand, the recognition process must be selective enough to be able to distinguish subtle details between action categories and on the other hand, recognition must be generalized across the actor's identity, size and position. The current study shows that the recognition of human actions is position sensitive, although the perceptive fields seem to be quite large and the perceptive field at fixation to a large degree invades the visual periphery and executes a large influence on the perception of actions at peripheral positions.

In contrast to the widespread view that along the cortical hierarchy object recognition is increasingly position invariant (DiCarlo & Cox, 2007; Hoffman & Logothetis, 2009; Riesenhuber & Poggio, 2000), other physiological studies already indicate position sensitivity at higher stages of processing for object recognition. It has been reported that even at processing stages involving IT position sensitivity can still be found. Single cell recording has revealed that receptive field sizes of neurons in IT range from 2.5° to 25° (Op de Beeck & Vogels, 2000). This already indicates a large variety of receptive field sizes at that stage of processing, which hinders a clear conclusion about position sensitivity. Kravitz and colleagues (2010) showed, with a behavioral object recognition experiment, significant reduction of priming effects with changes in position. Although the precise extent of this position sensitivity remains unclear, since most of the research has tested position shifts smaller than 7° (Riesenhuber and Poggio 1999; Kravitz, Kriegeskorte, & Baker, 2010; Sayres et al., 2015; Schwarzlose, Swisher, Dang, & Kanwisher, 2008), it is possible that object representations simply span larger portions of the visual field. This is similar to what we have shown here for the recognition of human actions. Although the central perceptive field already spans large parts of the visual field one cannot speak of position invariance as we show that at -20° eccentricity another action sensitive channel is being stimulated.

## 4.6 Conclusion

In the current study, we used an action adaptation paradigm to measure perceptive field sizes of action sensitive channels. The perceptive fields span over large portions of the visual field and decrease towards the periphery. The overlap of perceptive fields might seems to be leading to an influence of the action channel in central vision on the recognition of actions at adjacent positions in the visual periphery. The influence of the largest field in the fovea on peripheral positions might explain the high and nonlinearly declining recognition performance that we found in a previous experiment (Fademrecht et al., 2016).

# 5 STUDY IV: ACTION ADAPTATION IN A CROWDED ENVIRONMENT

Laura Fademrecht[1], Judith Nieuwenhuis[1], Isabelle Bülthoff[1], Nick Barraclough[2], Stephan de la Rosa[1]

[1] Max Planck Institute for Biological Cybernetics, Tübingen, Germany

[2] University of York, York, UK

## 5.1 Abstract

In real life, we need to recognize the actions of other individuals whether they are alone or surrounded by other people or when their actions are perceived outside of our foveal vision. Even though action recognition under such complex circumstances is crucial for social functioning, little is known about action recognition in such viewing conditions. In the current study, we therefore investigated whether the presence of a crowd has an impact on action recognition using an action adaptation paradigm. For higher ecological validity, we used life-size moving human figures to study the high-level visual mechanisms underlying action categorization. We assessed action recognition in two tasks (a recognition and an adaptation task) in central vision and at 40° eccentricity under four different viewing conditions: the moving figure was presented (1) alone, (2) in a crowd of static actors, (3), in a crowd of neutrally moving (idling) actors or (4) in a crowd of actors that performed the same actions as the adaptation stimulus. In both tasks we found recognition and adaptation performance was little affected by the crowd both at fixation and in the periphery with aftereffects larger at fixation. Our results suggest that action recognition mechanisms are robust even in visually distracting environments at fixation and in the periphery when tested under natural viewing conditions.

## 5.2 Introduction

Prolonged exposure to a visual stimulus, often called adaptation, can transiently change the subsequent percept of an ambiguous test stimulus away from the adapted stimulus. For example, after adapting to a red square a white square is perceived with a greenish tint. The presence of such adaptation aftereffects may be explained by shared visual processes between adaptor and test stimuli (e.g. the pooled response across several color channels). Adaptation is thought to alter the response properties of visual processes involved in the perception of the adaptor. If these processes are partially shared between adaptor and test, these alterations are passed on to the perception of the test stimulus thereby changing its percept (Webster 2011). By systematically varying the visual similarity between adaptor and test stimulus, adaptation aftereffects can be used to assess the tuning characteristics of visual processes. This method has therefore also been called the psychophysicist's microelectrode (Frisby, 1979).

Adaptation aftereffects have become a popular paradigm for behavioral study of the response properties of visual processes. While early work on visual adaptation aftereffects focused on low-level stimulus properties such as color, motion, and orientation (C.W.G. Clifford, 2002; Gibson & Radner, 1937; M. A. Webster & Leonard, 2008), in recent decades scientists have started exploring high-level adaptation aftereffects. Most of this research has focused on face perception. For example, reliable adaptation aftereffects have been demonstrated for the perception of facial characteristics, such as sex, attractiveness, ethnicity, and identity (Leopold et al., 2001; Rhodes et al., 2003, 2010; M. A. Webster et al., 2004). These studies suggest that adaptation is not a unique mechanism of the low-level sensory cortex, but can also target higher-level cortical areas.

Adaptation paradigms have also been applied to study the visual processes underlying action perception. Previous research mainly focused on investigating the visual mechanisms regarding the perception of gender (Jordan et al., 2006; Troje et al., 2006) and emotions (Roether et al., 2009) from human biological motion, walking direction discrimination (Barraclough & Jellema, 2011) and weight judgments with object-directed actions (Barraclough et al., 2009). However, action categorization, an essential process for social behavior, has received much less attention. Action categorization is important for social interactions as it enables the observer to choose an appropriate response to a variety of body movements. The visual mechanisms underlying action categorization are poorly understood. One of the first studies uniting an adaptation methodology with the study of high-level influences on categorical action perception was conducted by de la Rosa and colleagues (2014). They employed an action adaptation paradigm to show that action recognition is modulated by social context. Participants categorized static images of ambiguous actions that were rendered from a video showing the body posture transition between a wave and a punch. The authors found that action adaptation aftereffects were modulated by social action context (friendly or hostile) that preceded the action although the physical properties of adaptor and test stimuli were unchanged. These findings support the idea that action categorization is modulated by high-level influences.

To better understand action recognition in real-life situations, it is necessary to examine action recognition under more naturalistic viewing conditions than has previously been done. In real life, the observer is often required to recognize actions in the presence of other people and both in central vision and the visual periphery. Take for example a defender in a football match who is running towards an opponent who is in possession of the ball. In this situation the defender needs to recognize whether the opponent is passing the ball or is running with it. Importantly this task is done in a visual background crowded with other moving players. Moreover, the defender needs to monitor the actions of other opponents in his visual periphery that might try to join the attack. The current study sought to examine the influence of these two important factors on the categorization of actions: the presence of other people in the scene and the presence of the stimulus in the visual periphery instead of foveal vision.

Additional individuals in the visual field, standing close to the actor, could induce the well-known crowding effects, especially in the visual periphery. Crowding has a deleterious effect on visual recognition of objects and actions and is mostly found in peripheral vision, due to the decline of visual acuity towards the periphery (Levi, 2008). In action recognition, crowding has been studied in the context of biological motion with point-light walkers. In a direction discrimination task, Ikeda et al. (2013) showed that crowding occurred only with walking flankers but not with scrambled flankers, thus indicating that crowding of biological motion is a high-level effect. In the experiment of

Thornton and Vuong (2004) participants performed a flanker task and were asked to discriminate the walking direction of the central walker while ignoring the flankers. The results show that biological motion can be processed passively in a bottom-up fashion and therefore the flankers' walking direction influenced the perception of the target stimuli's walking direction. However, little is known about the effect of crowding on the human ability to distinguish different actions. Applying clutter in the form of additional actors in the visual scene allows us to investigate the degree to which 'natural visual clutter' in the scene negatively impacts the visual processes underlying the ability to tell different actions apart. To that end, we used an action recognition and action adaptation task. The latter allows selective targeting of action categorization processes.

Previous research on peripheral action recognition is scarce, especially concerning the far visual periphery. The few available studies used point-light walkers in detection and direction discrimination tasks at eccentricities up to only 12° away from fixation. Results show that such actions can be detected reliably in this "near" peripheral range, though performance is better in foveal vision (Ikeda et al., 2005, 2013; B. Thompson et al., 2007). In a recent study about action recognition in the far periphery, participants viewed human actions at different positions in the visual field (up to 60° eccentricity) and categorized their actions at various categorization levels (a handshake could be classified as a handshake or a greeting for example). This study showed that action categorization performance in the far periphery is comparable to central vision (Fademrecht, Bülthoff, and de la Rosa, 2016). Here, with a setup similar to the one used by Fademrecht and colleagues, we investigate action discrimination at 0° and 40° eccentricity. We assessed whether action discrimination performance at these locations is robust against the presence of a crowd in two different tasks, namely action adaptation (Experiment 1) and action recognition (Experiment 2).

In summary, the aim of the current study is to investigate action recognition under more realistic viewing conditions by examining the influence of a crowded visual environment on action adaptation aftereffects and action recognition performance both in the fovea and the visual periphery. In Experiment 1, we use an action adaptation paradigm similar to the one de la Rosa et al. (2014) developed to investigate action recognition processes.

## 5.3 Experiment 1

### 5.3.1 Methods

*Participants:* 28 participants (18 females) were recruited from the local community of Tübingen. Participants received monetary compensation for their participation in the experiment. Their ages ranged from 20 to 56 years (*M*: 28.5; *SD*: 8.5). Participants' visual acuity was normal or corrected to normal with contact lenses, since glasses might occlude parts of the visual periphery. Participants provided written informed consent prior to the experiment. The study was conducted in accordance with the Declaration of Helsinki and under advisement of the ethics board of the University of Tübingen.

*Apparatus:* A large panoramic screen with a semi-cylindrical 2D projection system was used for the presentation of the stimuli (Figure 8). The almost semi-circular screen is 3.2 m high and 7 m long. It covers 230° horizontally and 125° vertically of the visual field of our participants seated in front of the screen. The basic geometry of the screen can be described as a quarter-sphere. Six LED DLP projectors (1920x1200, 60Hz) (EYEVIS, Germany) were used to display the stimuli against a grey background on the screen. We used warping technology software (NVIDIA, Germany) to compensate for the visual distortions of the display caused by the curved projection screen. The stimuli had the size of a human figure placed 4 m to 6 m away from the participant. During experimental trials participants were required to focus their gaze on a white fixation cross presented on the screen straight ahead of them. The position of the cross was defined as the 0° position. The Unity 3D (Unity Technologies, USA) game engine in combination with a custom written control script was used to control presentation of the stimuli and response collection of keypresses given by the participants.

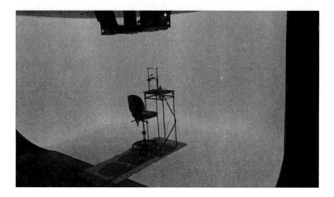

*Figure 8: Experimental Setup. Semi-cylindrical screen with the participant positioned in the center.*

*Stimuli:* Two human actions (hug and clap) were recorded from one actress (height: 168 cm) via motion capture using MVN Suits (XSens, Netherlands). The Xsens MVN Suit consists of 17 inertial and magnetic sensor modules placed in an elastic lycra suit worn by the actor. The sampling rate of the sensors was 120 Hz. Both actions started with a neutral body position and lasted 1385 ms. The actions ended in a peak frame, which was specified as the point in time just before the actor started moving back to the neutral position. The biological motion data was mapped onto a virtual avatar. To display the actions, we mapped the biological motion data onto a life-size grey stick figure avatar (height: 170 cm, approximately 24° visual angle (VA)). The choice of using a stick figure instead of a more realistic avatar prevented other visual cues like appearance or gender from influencing participants' decisions besides the motion of the body. The position of the stick figure avatar on the screen was defined by the position of a point midway between both hips. The avatar was always oriented towards the participant. Its position on the screen varied between 0° eccentricity and 40° eccentricity to the right.

To create ambiguous action stimuli a morphing algorithm was implemented to create body motions in between the hug and the clap actions. The weighted averages of the positions for each joint on the body (for example the elbow) for the two actions were calculated for each action frame. We adjusted the morph levels of the ambiguous test stimuli for each participant individually. In order to obtain five different morph levels that participants perceived to be ambiguous, we presented the morph between the two actions in 0.1 proportion steps and asked each participant to indicate when the morphed action looked ambiguous (stimuli were presented at 0° VA). For all participants five morph levels around their individual ambiguous morph value were chosen. They were equally spaced by a 0.025 morph level value (minimum morph level 0.33; maximum morph level 0.63).

Each action stimulus was presented on the screen in four different crowd conditions: no crowd, static crowd, neutral crowd and active crowd. In the no crowd condition, the target stimuli were presented alone on the screen. In the in the other crowd conditions, the target stimuli were surrounded by 16 additional stick-figures distributed pseudo-symmetrically left and right of the fixation position. We would like to point out that the crowd was presented during both the adaptation and test phases. The crowd avatars were positioned on an arc of a circle (6 m virtual distance away from the participant), with the whole crowd spanning 140° VA. Avatars were first distributed evenly in the crowd (separated by 9.33° VA). Then an unequal spacing between avatars was obtained by adding random jitter in the x- and z-coordinates (range: ⊠ 1.17° VA) along the arc of a circle to make the crowd's positioning appear more natural. These figures performed distinct idle movements continuously in the neutral crowd condition, or were fixed to the posture of the first frame of the idle movement sequence in the static crowd condition. Various idle movements (e.g. stepping from one foot to the other, shaking one leg) were selected from Rocketbox Libraries (Havok, Ireland) and applied to each figure randomly. Selection criteria for the idle animations were that the arms were never lifted above the chest, and that the animation was calm and moderately paced. These criteria were applied to allow clear distinction between the clap and hug actions and the idle animations, thus ensuring easy identification of the target stimulus. In the active crowd condition, the crowd members were randomly assigned with either the hug or the clap action.

The crowd members that were closest to the target stimuli were positioned at a distance of 4.7° VA ⬚ 1.17° VA from the adaptor and test stimuli. The shoulder width of the stick figures amounted to 6° VA for the target stimuli and between 3.6° VA and 5 ° VA for the crowd stick figures (due to the larger virtual distance of the crowd). When the jitter in x- and z-coordinates was maximal, the distance between the shoulders of the flankers and the target stimulus were 0.1° apart. During execution of the actions however, the arms of the stick figures were moving, which led to an overlap to a certain degree of the target stimuli and the crowd members in all the trials.

*Procedure and Design:* Participants were seated in the middle of the arena and their heads were stabilized with a chin and forehead rest placed on a desk in front of them (see Figure 1). A fixation cross was continuously present during target stimulus presentation. In the baseline condition we probed participants' perception of the test stimuli without prior adaptation both at 0° and 40° eccentricity. Each trial began with presentation of the fixation cross and, after 500 ms, the test stimulus for 1385 ms. After an inter stimulus interval (ISI) of 500 ms the question: "What did it look like?" and the response options "hug" and "clap" appeared on the screen. Participants were asked to respond using corresponding keys on a keyboard. Each of the five ambiguous test stimuli was presented three times in random order. Presentation location (0° vs 40° eccentricity) was also randomized.

After the baseline measurements, the experimental conditions were probed which tested all possible combinations of our experimental manipulations: the adaptation aftereffect was tested in three different crowd conditions (no crowd, static crowd and moving crowd) and at two different eccentricities (0°, 40° eccentricity), as described above. The hug and the clap actions served as adaptor stimuli while each of the five ambiguous morphs were used as test stimuli. In every experimental condition participants' task was to categorize the actions of the test stimulus as either clap or hug.

Adaptor conditions were blocked, such that participants first completed all trials either with the hug or the clap action as the adaptor stimulus. Eccentricity was also blocked, meaning that within each of the adaptor conditions, participants first completed all trials either with the adaptor and test stimulus presented at 0° eccentricity, or with the adaptor and test stimulus presented at 40° eccentricity. Adaptation and test always occurred at the same position. Finally, within each eccentricity block, the crowd conditions were blocked as well with a randomized presentation order of crowd blocks (no crowd, static crowd and neutral crowd and active crowd). Adaptor block order was pseudo-randomized across participants. We used a between-subjects design in the sense that 16 participants performed the no crowd, the static crowd and the neutral crowd condition and 12 participants performed the no crowd and the active crowd condition.

Figure 9 shows the chronological sequence of an experimental trial. Each first trial in an experimental condition started with an initial adaptation phase in which the adaptor stimulus was shown 26 times (inter stimulus interval (ISI) = 500ms). After the initial adaptation phase, the experimental trials were presented. Each trial consisted of the presentation of the adaptors (either hug as adaptor or clap as adaptor) repeated 4 times. Adaptation was followed by an audible beep sound (1000 Hz), then, after a 500 ms ISI, by one of the five ambiguous test stimuli and the answer screen (Figure 2). Participants had unlimited time to respond. The next trial started immediately after the participants gave their response via keypress. Participants were explicitly instructed that a decision about the category of the test stimulus (hug or clap) was expected only during the test phase, not during the adaptation phase. Participants were asked to report their subjective feeling without trying to be constant in their answer patterns. Within an experimental condition we probed action categorization for each of the 5 morph levels three times for a total of 45 trials per experimental condition. Test stimuli presentation was randomized.

An Eyelink II eye tracker mounted on the chin rest recorded participants' eye movements. Participants were informed to fixate the fixation cross because otherwise the data could not be used for analysis. We had planned to remove from analysis trials for which participants moved their gaze away from the fixation cross by more than 2° during the stimulus presentation. Due to a technical error, the eye-tracking data could not be used. However, previous research using the same testing environment has shown that participants can reliably fixate (proportion of invalid trials was 1%) even during stimulation of the visual periphery.

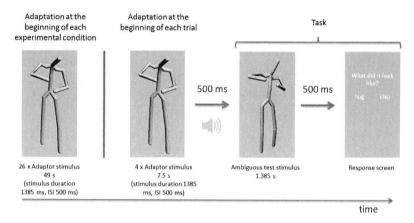

*Figure 9: Timeline of an experimental trial.*

## 5.3.2 Results

Two participants were excluded due to their performance in the baseline condition: they did not perceive the morphed test actions ambiguously and always gave the same response to all test stimuli.

We assessed the adaptation effect for each adaptor separately (hug and clap adaptation aftereffects). For all conditions, results were assessed in terms of proportion of clap responses. The adaptation aftereffect was obtained by subtracting the proportion of clap responses after exposure to an adaptor (hug or clap) from the proportion of clap responses in the condition without adaptation (baseline condition). We also calculated the overall change in perception (overall adaptation effect) as the difference in clap responses between the hug and clap adaptation conditions.

Figure 10 illustrates the hug and clap adaptation aftereffects in the four crowd conditions at fixation and in the periphery. As expected, we observed an antagonistic adaptation effect in all experimental conditions. Specifically, adapting to a hug action resulted in participants perceiving the ambiguous action more like a clap (increase of clap responses relative to baseline). Similarly, adapting to a clap action leads participants to perceive the test action as a hug (i.e. a decrease of the number of clap responses relative to baseline). Figure 11 shows the overall adaptation effect. Visual inspection of this figure suggests that adaptation effects are smaller for 40° eccentricity than for 0° eccentricity.

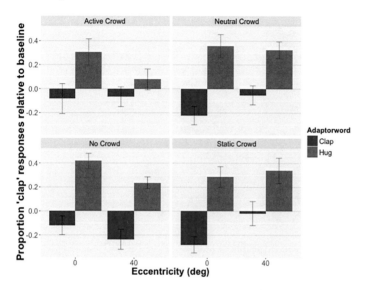

*Figure 10: Average difference in number of clap responses between baseline condition and adapted condition for each adaptor. Colors indicate which adaptor was used. Error bars represent standard error of the mean.*

To more formally assess the effects of the crowd condition and presentation eccentricity on the overall adaptation effect, a linear mixed model with crowd condition and eccentricity as fixed factors and participant as random factor was calculated. We allowed the intercept and the slope to vary randomly in a per participant fashion within both fixed factors. The results show that the overall adaptation aftereffect is significantly stronger at 0° eccentricity than at 40° ($F(1,65)$ = 12.78, $p$ = 0.02). There was no significant main effect of the crowd condition ($F(3,65)$ = 4.01, $p$ = 0.81). Hence, the static crowd, the neutral crowd and the active crowd had little influence on the adaptation aftereffect in comparison to the no crowd condition (Figure 4). The interaction between eccentricity and crowd condition was non-significant ($F(3,65) = 0.96$, $p = 0.22$).

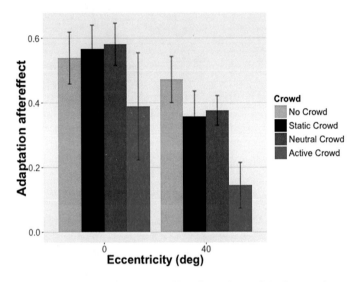

*Figure 11: Overall adaptation aftereffects for each crowd condition. Colors represent the four crowd conditions. Error bars represent standard errors of the mean.*

## 5.3.3 Discussion

In Experiment 1, we used an action adaptation paradigm to study the robustness of action recognition mechanisms to the presence of a crowd and to peripheral viewing cond3tions. We tested whether the presence of other humans in the visual scene influenced action adaptation aftereffects. We carried out the same tests at fixation and in the far periphery (40° eccentricity) to test peripheral viewing. Our study reveals a clear and robust adaptation aftereffect on the perception of actions at both eccentricities and with all crowd conditions. At the same time the adaptation effect demonstrates that the perception of actions is malleable: it can be altered by the prior prolonged presentation of another action even in the far periphery. Antagonistic adaptation effects (i.e. perceptual biases away from the adaptor action) have often been interpreted in terms of a contrastive organization of the underlying visual processes (for details see Leopold et al., 2001; Webster, 2011). In line with Leopold et al. (2001) and Webster (2011), we suggest that high-level action adaptation aftereffects might be explained in terms of perceptual channels sensitive to actions (see Gardner, 1973 for more details about perceptual channels). Our results suggest that the contrastive representation holds true for the two actions used in the experiment, which can be seen in the antagonistic effect of the two adaptation conditions. We thereby extend previous research of movement direction adaptation (Barraclough and Jellema, 2011) into the realm of social action adaptation and propose that perceptual channels concerned with the discrimination of actions are organized in a contrastive fashion. The contrastive representation was also shown in other studies in our lab for different action pairs, for example punch and handshake, punch and fist-bump, handshake and high-five (see also de la Rosa et al. 2014). Whether this is generalizable to all possible human actions is not clear and would be subject to future research.

The adaptation aftereffect was significantly reduced in the periphery compared to fixation. This reduction at 40° eccentricity could originate from the decreased visual resolution found at such a far eccentricity. Indeed, previous research on biological motion stimuli has shown that biological motion perception in the periphery suffers in comparison to central vision (Ikeda et al., 2005, 2013; B. Thompson et al., 2007). Nevertheless, the mere presence of an adaptation aftereffect even at 40° eccentricity indicates that participants were able to recognize the action even that far into the visual periphery. This is broadly in line with other research that suggests that participants are remarkably good at recognizing actions in the visual periphery (Fademrecht, Bülthoff, and de la Rosa, accepted). Moreover, the fact that antagonistic adaptation aftereffects were observed even in the visual periphery suggests that contrastive encoding of actions is not specific to foveal vision.

The non-significant effect of crowd condition could be related to previous research that has shown that other adaptation effects (e.g. orientation adaptation) are little affected by crowding. For example, Blake et al. (2006) showed that crowding does not reduce the adaptation aftereffect for simple features (e.g., orientation-dependent threshold-elevation aftereffect) when stimuli are presented with high contrast. Similarly, Pelli and

Tillman (2008) report that crowding prevents the ability to judge target orientation, while it permits the occurrence of an orientation specific adaptation aftereffect. While these studies report that orientation adaptation is little affected by crowding, as we found in our own study with action adaptation, the suggested underlying mechanism might not apply to the results of our study. Specifically, Pelli and Tillman (2008) explain these effects within a two-step object recognition processes. In the first step, which is susceptible to adaptation, object features are detected. In the second step features are combined. According to Pelli and Tillman (2008), feature combination is susceptible to crowding. This explanation is more difficult to reconcile with the results of our study because the visual features critical for the recognition of actions are assumed to be combinations of 'object' features of Pelli and Tillman's first stage (Pelli & Tillman, 2008). Hence according to Pelli and Tillman's explanation, one would expect action adaptation to be affected by crowding, which is not what we found. Based on our findings we do not want to exclude the possibility that crowding might affect action adaptation for other crowd configurations. For example, our crowd stimuli were presented at a larger virtual distance from the observer than the target stimuli (6 m vs 4 m away). Our study contains a variety of monocular depth cues (e.g. occlusion, vertical position in the field, relative size), do those monocular depth cues influence crowding in our experiment? Freeman and Simoncelli (2011; Whitney and Levi 2011) have shown that crowding occurs with images containing monocular depth cues suggesting that these cues have only little effects on crowding.

In order to examine whether the results of Experiment 1 generalize to other recognition task, we probed the sensitivity of action recognition to peripheral and foveal presentation of actions under different crowd conditions in Experiment 2.

## 5.4 Experiment 2

### 5.4.1 Methods

We used the same methods as in Experiment 1 except for the following:

*Participants:* We recruited 17 participants (thirteen female) from the local community of Tübingen (of which 12 also participated in the first experiment). The ages ranged from 21 to 56 (*M*=31.29, *SD*=9.82).

*Procedure and Design*: The clap action and the hug action (no morphed actions were used) were presented at 0° and 40° eccentricity randomly in all conditions (no crowd, neutral crowd and active crowd condition, there was no static crowd condition). Participants were asked to categorize each action as either clap or hug as fast and accurately as possible. Answers could be given any time directly after stimulus onset via key-press. As measures of recognition performance participants' accuracy and reaction times were recorded.

## 5.4.2 Results

We assessed recognition performance in terms of reaction time and accuracy. The results for the two dependent variables are presented separately. Only reaction times for correct responses were considered in this analysis. Participants' reaction times did not increase at 40° eccentricity and did not depend on the crowd condition (Figure 12). Moreover, there seems to be some modulation in reaction times with crowd conditions at foveal presentations but no modulation of reaction times with crowd condition at 40° eccentricity. A linear mixed model with crowd condition and eccentricity as fixed factors and participant as random factor was calculated. For reaction times (Figure 23) we found a significant main effect of eccentricity ($F(1, 80) = 13.919$, $p < .001$). The main effect of crowd ($F(2, 80) = 0.594$, $p = 0.555$) and the interaction between crowd and eccentricity ($F(2, 80) = 2.397$, $p = 0.098$) were non significant.

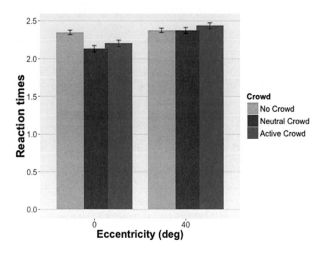

*Figure 12: Mean reaction times and standard errors for the three crowd conditions at 0° and 40° eccentricity*

The mean accuracy was 94 % ($SE$ = 0.3%) and therefore well above chance level for both tested eccentricities and the three crowd conditions (Figure 13). The accuracy results were analyzed using a linear mixed model with crowd condition and eccentricity as fixed factors and participant as random factor. The results show a significant main effect of eccentricity ($F(1, 80)$ = 25.245, p < .001), while the main effect of crowd ($F(2, 80)$ = 0.002, p = 0.998) and the interaction of crowd and eccentricity ($F(2, 80)$ = 0.003, p = 0.999) were non significant.

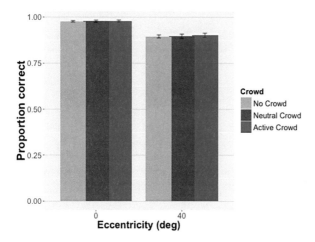

*Figure 13: Mean proportion correct and standard errors for the three crowd conditions at 0° and 40° eccentricity*

### 5.4.3 Discussion

The results reveal a significant increase in reaction time in far visual periphery compared to central vision suggesting that recognition performance is worse in the periphery. This result is in line with previous findings of Fademrecht and colleagues (2016), showing that action recognition is possible in far periphery but decreases for eccentricities larger than 30°. The presence of a crowd had no significant influence on the participants' action recognition performance. Even in the visual periphery where crowding effects could be anticipated the presence of the crowd had no significant influence on the recognition performance. This result is surprising in the light of other findings. For example, Ikeda and colleagues (2013) as well as Thornton and Vuong (2004) showed effects impairing effects of flanking stimuli on the perception of point-light walkers. Our experiment differs in several ways from these previous studies. First, we used a stick figure avatar instead of the point-light stimuli employed in previous research. The stick figure provides more form information than point-light stimuli, which might enhance recognition performance. Second, our stimuli were life sized while previous stimuli were much smaller. It is possible that recognition of actions in crowded environments perform better under these more natural stimulus sizes conditions. In any case our results demonstrate that the presence of a crowd has only little influence on the action recognition when action recognition is assessed in more realistic viewing scenarios with life size human-like actors.

## 5.5 General Discussion

Embedding an action into a crowd did not significantly modify action adaptation effects or action recognition performance. Surprisingly, neither a crowd, performing neutral idle movements that did not resemble the target actions, nor a crowd of people that performs the actual target actions does influence human action recognition and action adaptation aftereffects. These findings imply that action-sensitive visual processes are little influenced by the presence of other individuals in the scene. Such robustness of visual categorization mechanisms with regards to crowded environments would clearly be advantageous for real-life social interactions where actions often occur in the presence of other people.

To what degree can additional individuals in the visual field, standing close to the actor, induce crowding effects in our experiments? When relating our results to findings in the crowding literature, it is important to note some differences with regards to stimulus contrast stimulus size. For example, our stimuli were shown at a contrast level and size level well above detection threshold level (i.e. supra-threshold). What is the effect of supra-threshold size and contrast on crowding effects? We argue that crowding effects can be observed with supra-contrast threshold stimuli, as Ikeda et al. (2013) have shown in their study about crowding in biological motion. Specifically, Ikeda et al. (2013) used supra-contrast threshold stimuli and were able to observe crowding effects for biological motion stimuli that were 4° tall with maximum contrast (white point-light display on black background) presented at 5° eccentricity. In their study, the stimulus contrast was much higher than in our experiment which presented grey stick figures on a grey background. In terms of contrast our stimuli are less supra-threshold than in the study of Ikeda et al. (2013) while being still highly visible and leading to a high recognition performance as the results of Experiment 2 show. Although the size of our stimuli was much larger than in the study of Ikeda et al. (2013), we assessed adaptation aftereffect and action recognition at much larger eccentricity. Critical for the occurrence of crowding effects is usually the critical distance between the target and the flankers. Crowding happens when the target-flanker separation is smaller than a critical distance. Bouma (1970) outlines an estimate for the critical distance for which crowding can be observed as half the eccentricity of the stimulus' presentation location. In our study, the stimuli are presented at 0° and 40° eccentricity, which results in a critical spacing of 0° in the fovea and a critical spacing of 20° at 40° eccentricity. Because the test stimuli overlapped with stick figures of the crowd on every trial, the target-flanker separation was sufficiently small for crowding to occur. In sum, the visual parameters for our stimuli are likely to have allowed for crowding effects to occur in our experiments despite the supra-threshold nature of the stimuli. Our results indicate that action adaptation aftereffects and action recognition very little affected by crowding.

An important aim of the present study was to use a paradigm with increased ecological validity compared to previous research in this field. To this end we chose life-size human stick figures that performed actions as stimuli. Although the resemblance between stick figures and real-life actors may be disputed, we argue here that the benefits outweigh the costs: we were not only able to use dynamic actions rather than static images thereof, but we furthermore minimized the chances that our results are clothing-, body shape-, or gender-specific. Despite their simplicity, stick figures provide a clear step toward real-life actors when compared to the point-light walkers used in many previous studies. Additionally, by the use of a panoramic display we were able to test a large portion of the horizontal extent of the visual field, which increased the generality of our finding as well and moves this research towards real-life conditions.

## 5.6 Conclusion

Action adaptation is a useful tool to specifically target action recognition processes. Using action adaptation, we observed a significant adaptation aftereffect at fixation as well as in the visual periphery, although the latter was slightly weaker. It suggests that the human ability to categorize human social actions is extremely robust across the visual field. Furthermore, using an adaptation paradigm and an action recognition task we showed that the presence of a crowd does not impair action discrimination in both cases, in central vision and in far periphery.